AF556850

HUMAN ANATOMY AND PHYSIOLOGY

# MUSCLE CELLS

# DEVELOPMENT, DISORDERS AND REGENERATION

# HUMAN ANATOMY AND PHYSIOLOGY

Additional books in this series can be found on Nova's website under the Series tab.

Additional E-books in this series can be found on Nova's website under the E-book tab.

HUMAN ANATOMY AND PHYSIOLOGY

# MUSCLE CELLS

## DEVELOPMENT, DISORDERS AND REGENERATION

BENIGNO PEZZO
EDITOR

*New York*

For permission to use material from this book please contact us:
Telephone 631-231-7269; Fax 631-231-8175
Web Site: http://www.novapublishers.com



**Library of Congress Cataloging-in-Publication Data**

ISBN: 978-1-62417-233-5

Library of Congress Control Number: 2012950969

*Published by Nova Science Publishers, Inc. † New York*

# CONTENTS

# PREFACE

This book discusses current research in the study of the development, disorders and regeneration of muscle cells. Topics include the basic biology and current concepts of muscle regeneration; calvarial and periodontal tissue induction by autogenous striated muscle stem cells; the role of muscle cells on the processus vaginalis; a discussion on the key residues which cause differential gallbladder response to PACAP and VIP in the guinea pig; and controlling muscle functions with light.

Chapter 1 – Beyond skeletal muscle's primary function as a force generator for locomotion, there is a growing recognition of the important role skeletal muscle plays in overall health through its impact on whole-body metabolism as well as directly influencing quality of life issues with chronic disease and aging. Over the last decade, extensive progress has been made with regard to our understanding of the molecules that regulate skeletal muscle regeneration. Satellite cells are muscle-specific stem cells located under the basal lamina of muscle fibers, which are responsible for muscle regeneration. Similarly to the embryonic stem cells that build organs, adult stem cells that regenerate organs are capable of symmetric and asymmetric division, self-renewal, and differentiation. This precise coordination of complex stem cell responses throughout adult life is regulated by evolutionally conserved signaling networks that cooperatively direct and control (1) the breakage of stem cell quiescence, (2) cell proliferation and self-renewal, (3) cell expansion and prevention of premature differentiation and finally, (4) the acquisition of terminal cell fate. This highly regulated process of tissue regeneration recapitulates embryogenic organogenesis with respect to the involvement of interactive signal transduction networks. Indeed, various modulators such as insulin-like growth factor-I (IGF-I), hepatocyte growth factor (HGF), leukemia inhibitory factor (LIF), and Wnt have been shown to stimulate the activation and proliferation of satellite cells. PI3-K (phosphatidylinositol 3-kinase)/Akt/mTOR (mammalian target of rapamycin), calcineurin, and serum response factor (SRF) seem to contribute to muscle regeneration by regulating differentiation of satellite cells in co-operation with the MyoD family and/or p21. In contrast, myostatin inhibits these processes through forkhead box O (FOXO) and/or Smad 2/3-dependent signaling. Various studies using *in vitro* cell cultures and *in vivo* rodent models have revealed candidates for proteins that modulate the regenerating process in muscle fibers after damage. In this chapter, the authors describe the molecular and cellular mechanisms regulating muscle regeneration.

Chapter 2 – The central question in developmental biology, tissue engineering and regenerative medicine at large, is the molecular basis of pattern formation, tissue induction and morphogenesis. The three requirements for the induction of tissue morphogenesis are a suitable biomimetic extracellular matrix substratum, soluble inductive molecular signals, and responding stem cells capable of ligating soluble molecular signals. Tissue induction and morphogenesis by combinatorial molecular protocols is epitomized by the sequential cascade of "*Bone: Formation by autoinduction*". Any of the three variables in the equation can be modified and modulated to initiate the induction of bone formation in skeletal defects of the craniofacial and appendicular skeletons. A number of isoforms of soluble osteogenic molecular signals may be recombined or reconstituted with different extracellular matrix substrata to biomimetize the structure/activity profile of the extracellular matrix as well as of the osteogenic soluble molecular signals. Stem cells with selected ligands' receptors are capable of differentiating and inducing selected tissue phenotypes and morphogenesis. Progenitor stem cells are either locally stimulated by available soluble molecular signals or can be additionally isolated and intra-operatively added to the surgical site providing an adjunctive tool to therapeutic bone tissue engineering. Striated muscle represents an abundant source of easily accessible tissue that contains several perivascular and intramuscular cell niches available for tissue engineering applications. Myoblastic stem cells including myoendothelial stem cells harvested from striated muscle represent a therapeutic advancement in regenerative medicine and tissue engineering for craniofacial and periodontal applications. Muscular tissue also contains mesenchymal stem cells now known to be pericytes attached to perivascular niches. Morcellated fragments of autogenous *rectus abdominis* muscle containing large quantities of pericytes and myoendothelial cells delivered by Matrigel® matrix and insoluble collagenous bone matrix recombined with recombinant human transforming growth factor-$\beta_3$ (hTGF-$\beta_3$) enhance calvarial regeneration in the non-human primate *Papio ursinus*. Morcellated fragments of autogenous *rectus abdominis* muscle combined with soluble osteogenic molecular signals induce greater amounts of alveolar bone and cementum regeneration along the exposed root surfaces in periodontal defects of *Papio ursinus*. Importantly, morcellated fragments of striated muscle are relatively surgically accessible not only from the *rectus abdominis* but from the orofacial muscular tissues. Harvested fragments require minimal surgical preparation and none *in vitro*, yet retain significant regenerative potential directed by the surrounding extracellular matrices, i.e. osteogenic in craniofacial osseous sites and cementogenic when in contact with dentine extracellular matrices.

Chapter 3 – Congenital inguinal hernia (IH) and hydrocele are among the commonest pathologies affecting children and both are caused by the incomplete obliteration of the processus vaginalis (PV) which normaly obliterates near the end of the gestational period or sortly after.

A number of factors, endocrine, neurophysiologic, cytologic, regulate PV development. These regulatory factors are not mutually exclusive in their action and the authors think that an experimental or observational finding that may affect the fate of the PV does not necessarily invalidate a seemingly contradictory theory based on other findings.

The normal process of PV obliteration is considered, by some authors, to include a stage of dedifferentiation of smooth muscle cells (SMCs) that are found on the PV, and their eventual apoptosis. Histological studies reveal the existence of SMCs on the wall of unobliterated PV. Sympathetic and parasympathetic nerve action, which in its turn is affected by hormones, is probably involved to produce or to halt such a result.

In this particular study the authors review the literature on these biologic mechanisms, including their own contribution which is the following: By using immuno-histochemical studies the authors examined the cytoskeletal proteins of SMCs present in the PV of patients with IH and hydrocele and drew conclusions on the degree of SMC dedifferentiation. Sacs from patients with IH and especially from male IH, have fully differentiated SMCs while sacs obtained from hydroceles are in an intermediate state of dedifferentiation. The authors' findings are suggestive that in cases of IHs the SMCs on the wall of the hernia sac do not follow the natural way of dedifferentiation and apoptosis, and only partly do so in cases of hydrocele. This may be the reason for the varying degree of incomplete obliteration of the PV in these cases.

Chapter 4 – The aim of this chapter is to investigate the effects of pituitary adenylate cyclase activating polypeptide (PACAP) and vasoactive intestinal peptide (VIP) in the guinea pig gallbladder, and identify key residues responsible for their interactions with PACAP (PAC1) and VIP (VPAC) receptors in the guinea pig gallbladder.

Chapter 5 – Traditionally, artificial contractions of muscles have been induced electrically, mechanically or pharmacologically to investigate their functional characteristics. Although simple and convenient, these techniques are generally non-specific, non-uniform and invasive. To improve the spatiotemporal resolution and to reduce the invasiveness, the optogenetic approach using light-sensitive proteins has attracted attention as a new method. Recent examples include using channelrhodopsin-2 (ChR2), a light-activated ion channel from a green alga, for optical pacing of cardiomyocytes, the optical control of C2C12 myoblast-derived myotubes and the optically induced maturation of cultured myotubes. The optical manipulation of muscle activities would facilitate *in vitro* studies of muscle contraction through manipulating/modulating specific biological processes during myogenic development. It has potential therapeutic applications for producing light-sensitive human muscle substitutes for muscle weakness such as muscular dystrophy and amyotrophic lateral sclerosis (ALS). It could also enable the development of a wireless driving source of muscle-powered actuators/microdevices. Here, this chapter reviews a general overview of the state of research and future prospects and challenges of optogenetics for muscle cells.

In: Muscle Cells
Editor: Benigno Pezzo

ISBN: 978-1-62417-233-5


*Chapter 1*

# BASIC BIOLOGY AND CURRENT CONCEPTS OF MUSCLE REGENERATION

***Kunihiro Sakuma**[*1] **and Akihiko Yamaguchi**[2]*

[1]Research Center for Physical Fitness, Sports and Health, Toyohashi University of Technology, Tenpaku-cho, Toyohashi, Japan

[2]School of Dentistry, Health Sciences University of Hokkaido, Kanazawa, Ishikari-Tobetsu, Hokkaido, Japan

## ABSTRACT

Beyond skeletal muscle's primary function as a force generator for locomotion, there is a growing recognition of the important role skeletal muscle plays in overall health through its impact on whole-body metabolism as well as directly influencing quality of life issues with chronic disease and aging. Over the last decade, extensive progress has been made with regard to our understanding of the molecules that regulate skeletal muscle regeneration. Satellite cells are muscle-specific stem cells located under the basal lamina of muscle fibers, which are responsible for muscle regeneration. Similarly to the embryonic stem cells that build organs, adult stem cells that regenerate organs are capable of symmetric and asymmetric division, self-renewal, and differentiation. This precise coordination of complex stem cell responses throughout

[*] Address correspondence and reprint requests to: Kunihiro Sakuma, Ph.D. Research Center for Physical Fitness, Sports and Health, Toyohashi University of Technology, 1-1 Hibarigaoka, Tenpaku-cho, Toyohashi 441-8580, Japan, E-mail: ksakuma@las.tut.ac.jp, Fax: 81-532-44-6947

adult life is regulated by evolutionally conserved signaling networks that cooperatively direct and control (1) the breakage of stem cell quiescence, (2) cell proliferation and self-renewal, (3) cell expansion and prevention of premature differentiation and finally, (4) the acquisition of terminal cell fate. This highly regulated process of tissue regeneration recapitulates embryogenic organogenesis with respect to the involvement of interactive signal transduction networks. Indeed, various modulators such as insulin-like growth factor-I (IGF-I), hepatocyte growth factor (HGF), leukemia inhibitory factor (LIF), and Wnt have been shown to stimulate the activation and proliferation of satellite cells. PI3-K (phosphatidylinositol 3-kinase)/Akt/mTOR (mammalian target of rapamycin), calcineurin, and serum response factor (SRF) seem to contribute to muscle regeneration by regulating differentiation of satellite cells in co-operation with the MyoD family and/or p21. In contrast, myostatin inhibits these processes through forkhead box O (FOXO) and/or Smad 2/3-dependent signaling. Various studies using *in vitro* cell cultures and *in vivo* rodent models have revealed candidates for proteins that modulate the regenerating process in muscle fibers after damage. In this chapter, we describe the molecular and cellular mechanisms regulating muscle regeneration.

**Keywords**: skeletal muscle, regeneration, serum response factor, tumor necrosis factor-alpha, calcineurin

## ABBREVIATIONS

ActRIIB = activin receptor IIB
CCL = chemokine (C-C motif) ligand
CCR = CC-chemokine receptor
CDK = cyclin dependent kinase
CsA = cyclosporine A
CTX = cardiotoxin
COX-2 = cyclooxygenase-2
FGF = fibroblast growth factor
Fn14 = fibroblast growth factor-inducible 14
FOXO = forkhead box O
HDAC = histone deacetylase
HGF = hepatocyte growth factor
IGF-I = insulin-like growth factor-I
IKK = IκB kinase
IL = interleukin

JAK1 = Janus kinase 1
JNK = c-Jun N-terminal kinase
LIF = leukemia inhibitory factor
MAPK = mitogen-activated protein kinase
MEF2 = myocyte enhancer factor 2
MHC = myosin heavy chain
MRTF = myocardin-related transcription facotr
mTOR = mammalian target of rapamycin
mTORC = mTOR signaling complex
MuRF1 = muscle ring finger-1
NFAT = nuclear factor activated T-cells
NF-κB = nuclear factor-kappaB
NO = nitric oxide
NOS = nitric oxide synthase
PAI-1 = plasminogen activator inhibitor-1
PI3-K = phosphatidylinositol 3-kinase
PKC = protein kinase C
SRF = serum response factor
STARS = striated muscle activators of Rho signaling
STAT = signal transducer and activator transcription
TGF-β = transforming growth factor-β
TNF-α = tumor necrosis factor-α
TWEAK = TNF-like weak inducer of apoptosis
uPA = urokinase-type plasminogen activator
YY1 = Yin-Yang1

# 1. Introduction

Skeletal muscle contractions power human body movements and are essential for maintaining stability. Skeletal muscle tissue accounts for almost half of the human body mass and, in addition to its power-generating role, is a crucial factor in maintaining homeostasis. Given its central role in human mobility and metabolic function, any deterioration in the contractile, material, and metabolic properties of skeletal muscle has an extremely important effect on human health.

Several possible mechanisms for age-related muscle atrophy have been described; however the precise contribution of each is unknown. Age-related muscle loss is a result of reductions in the size and number of muscle fibers

[Lexell 1993] possibly due to a multi-factoral process that involves physical activity, nutritional intake, oxidative stress, and hormonal changes [Roubenoff and Hughes 2000; Sakuma and Yamaguchi 2012; Scott et al., 2010]. The specific contribution of each of these factors is unknown but there is emerging evidence that the disruption of several positive regulators [Akt and serum response factor (SRF)] of muscle hypertrophy with age is an important feature in the progression of sarcopenia [Sakuma et al., 2008; Sakuma and Yamaguchi 2010; Sakuma and Yamaguchi 2011a]. In addition, sarcopenia seems to include the defect of muscle regeneration probably due to the repetitive muscular damage. Indeed, the group of Conboy [Carlson et al., 2008; Conboy et al., 2003; Conboy et al., 2005] indicates that Notch-dependent signaling is impaired in sarcopenic muscle.

Upon tissue injury, the cues released by the inflammatory component of the regenerative environment instruct somatic stem cells to repair the damaged area [Stoick-Cooper et al., 2007]. The elucidation of the molecular events underpinning the interplay between the inflammatory infiltrate and tissue progenitors is crucial to devise new strategies toward implementing regeneration of diseased or injured tissues. Regeneration of diseased muscles relies on muscle stem cells (satellite cells) located under the basal lamina of muscle fibers [Mauro 1961], which are activated in response to cytokines and growth factors [Kuang and Rudnicki 2008]. The current lack of knowledge of how external cues coordinate gene expression in these cells precludes their selective manipulation through pharmacological interventions.

The inflammatory infiltrate is a transient, yet essential, component of the satellite cell niche and provides the source of locally released cytokines, such as interleukin (IL)-1, IL-6, and tumor necrosis factor-α (TNF-α), which regulate muscle regeneration [Kuang et al., 2008]. As an inducible element of the satellite cell niche, the inflammatory infiltrate provides an ideal target for selective interventions aimed at manipulating muscle regeneration [Peterson and Guttridge 2008]. However, because local inflammation regulates multiple events within the regeneration process, global anti-inflammatory interventions have both positive and negative effects on satellite cells [Mozzetta et al., 2009]. Thus, it is important to elucidate the intracellular signaling by which inflammatory cytokines deliver information to individual genes in satellite cells.

Similarly to the embryonic stem cells that build organs, adult stem cells that regenerate organs are capable of symmetric and asymmetric division, self-renewal, and differentiation. This precise coordination of complex stem cell responses throughout adult life is regulated by evolutionally conserved signaling networks that cooperatively direct and control (1) the breakage of stem cell quiescence, (2) cell proliferation and self-renewal, (3) cell expansion and prevention of premature differentiation and finally, (4) the acquisition of terminal cell fate. This highly regulated process of tissue regeneration recapitulates embryogenic organogenesis with respect to the involvement of interactive signal transduction networks such as hepatocyte growth factor (HGF), Notch, MyoD, calcineurin, and SRF [Al-Shanti and Stewart 2009; Mantovani et al., 2007]. This review aims to outline the molecular and cellular mechanisms of muscle regeneration.

## 2. Controlling the Immune Response

Acute muscle injuries initiate a predictable series of responses by specific myeloid cell populations. As in other tissues, Ly6C+/F4/80− neutrophils are the first responders and begin to appear at elevated numbers within 2 hours of muscle damage, typically peaking in concentration between 6 and 24 hours postinjury and then rapidly declining in numbers. Following the onset of neutrophil invasion, phagocytic macrophages begin to invade, reaching significantly elevated concentrations at about 24 hours post injury and continuing to increase in number until about 2 days post injury, when their numbers begin to decline sharply [Frenette et al., 2000; Ochoa et al., 2007]. Their invasion precedes an increase in the population of nonphagocytic macrophages that reaches a peak in the muscle at about 4 days post injury but remains significant for many days.

Skeletal muscle, like other tissues, initially responds to injury with an innate immune response driven by Th1 cytokines. Cytokines expressed during Th1-driven inflammatory responses, especially interferon-γ and TNF-α, drive the classical activation of macrophages forward an M1 phenotype, a proinflammatory population capable of perpetuating the inflammatory response [Gordon and Taylor 2005]. M1 macrophages can also promote muscle damage by the production of cytotoxic levels of nitric oxide (NO) generated by inducible NO synthase (iNOS) [Villalta et al., 2009]. M1 macrophages express $CD68^{+}$, which is a valuable, macrophage-specific marker for the M1 phenotype. Although CD206-expressing M2 macrophages can also express CD68 under some conditions [Linehan 2005], this coexpression likely highlights the phenotypic and functional plasticity displayed by macrophages present in inflammatory microenvironments [Mosser and Edwards 2008]. CD68, also called macrosialin or ED1 antigen [Dijkstra et al., 1985], is a receptor for oxidized low-density lipoproteins and CD68 ligation of oxidized lipoproteins can activate phagocytosis by M1 macrophages and increase their production of proinflammatory cytokines [Ottnad et al., 1995].

After M1 macrophages reach their peak concentration in injured and regenerative muscle, they are replaced by a population of M2 macrophages that can attenuate the inflammatory response and promote tissue repair. M2 macrophages are activated by Th2 cytokines: IL-4, IL-10, and IL-13 play particularly well-characterized roles in their activation [Gordon 2003]. M2 macrophages are a complex phenotype that has been divided into three subcategories that reflect functional and molecular specializations [Mantovani et al., 2004]. M2a macrophages are activated by IL-4 and IL-13 and can promote wound healing and tissue repair. M2b macrophages are activated by immune complexes or Toll-like receptors and release Th2, anti-inflammatory cytokines. M2c macrophages are activated by IL-10 and release cytokines that deactivate the M1 phenotype and can promote the proliferation of nonmyeloid cells.

M2-macrophage-specific CD antigens have now been associated with functions that are important in regulating macrophage activity and phenotype. For example, CD163 is a macrophage-specific receptor for complexes of hemoglobin and haptaglobin [Kristiansen et al., 2001] and ligation of CD163 can contribute to regulating macrophage phenotype by increasing the expression of anti-inflammatory cytokines especially IL-10 [Philippidis et al., 2004]. Furthermore, internalization and breakdown of the hemoglobin/haptoglobin complex can help to return extracellular hemoglobin concentrations to nontoxic levels, thereby reducing cellular damage following injury [Moestrup and Moller 2004]. Hemoglobin internalization and breakdown can also inhibit the production of cytolytic, free radicals by neurotrophils and M1 macrophages. Thus, CD163 ligation may contribute substantially to shifting macrophages from a M1 phenotype to an M2c phenotype, and it can reduce muscle damage mediated by free radicals.

## 2.1. The Functional Role of Macrophages during Muscle Regeneration

Recognition and phagocytosis of muscle cell debris are probably critical. Indeed while M1 macrophages enhance the proliferation of local myogenic precursor cells, M2 macrophages stimulate their fusion and differentiation [Arnold et al., 2007]. M1 macrophages release a complex milieu of prostaglandins, cytokines and chemokines, which have been implicated in promoting muscle precursor proliferation and transition to the differentiation stage. Chen et al. [2005] and Warren et al. [2002] have demonstrated the importance of TNF-α in promoting satellite cell proliferation during the early stages of muscle repair, whereas others have demonstrated its function as a chemoattractant for myoblasts and satellite cells *in vitro* [Al-Shanti et al., 2008; Lolmede et al., 2009]. IL-6 has also been revealed to play a role in progenitor cell recruitment [Al-Shanti et al., 2008; Wang et al., 2008], whereas elimination of IL-6 greatly diminishes muscle growth [Serrano et al., 2008]. In contrast, TNF-α and IL-6 have been shown to inhibit the differentiation and maturation of myoblasts suggesting that the transition from a pro-inflammatory M1 response to a tissue remodeling M2 response is essential for the progression of myogenic differentiation and muscle repair [Tsujinaka et al., 1996].

The shift in phenotype from M1 to M2 macrophages coincides with the beginning of myogenic differentiation in muscle progenitor cells [St Pierre and Tidball 1994]. Tidball and Wehling-Henricks [2007] have shown that the depletion of M2 macrophages severely disrupts myoblast differentiation and fusion resulting in decreased muscle fiber diameters. The release of IL-10, a characteristic marker of M2 macrophages, is thought to play a key role in promoting the fusion and maturation of myotubes [Arnold et al., 2007; Strle et al., 2007]. The administration of anti-inflammatory medication following acute injury, particularly cyclooxygenase-2 (COX-2) inhibitors, can markedly delay the muscle repair process [Mackey et al., 2007]. A careful balance and control of the macrophage phenotype, particularly the promotion of an M2 phenotype, has been suggested as a potential therapeutic strategy to promote in situ muscle repair.

Some molecular interactions are required for macrophage recruitment and function in damaged muscles. The muscle tissue of mice with a null mutation of CC-chemokine receptor (CCR)2, the CCL [chemokine (C-C) motif ligand] 2 receptor, undergoes regenerating defects including fibrosis and calcification after muscle damage. In addition, uPA (urokinase-type plasminogen activator)$^{-/-}$ macrophages fail to infiltrate damaged muscle [Bryer et al., 2008]. This failure is associated with defective muscle regeneration, demonstrating that uPA is required for the homeostatic response to injury. Mice lacking an inhibitor of uPA, PAI-1 (plasminogen activator inhibitor 1), exhibit increased uPA activity: injured muscle of PAI-1$^{-/-}$ mice shows evidence of increased macrophage accumulation, and of accelerated muscle repair [Koh et al., 2005]. Expression of uPA is apparently required for the expression of insulin-like growth factor-I (IGF-I). IGF-I suppresses the expression and activity of macrophage migration inhibitory factor and the transcription factor NF-κB (nuclear factor-kappaB), possibly directly regulating the persistence of inflammatory responses [Palumbo et al., 2007; Pelosi et al., 2007].

## 3. Hepatocyte Growth Factor and Neuronal Nitric Oxide Synthase

By 24 hours after muscle injury, satellite cells enter the G1/S phase of the cell cycle [Hawke and Garry 2001]. Two factors have been demonstrated to activate quiescent satellite cells. The first is HGF. Early experiments using single muscle fibers with associated quiescent satellite cells have shown that growth factors, such as IGF-I and fibroblast growth factor (FGFs), do not activate satellite cells in fibers [Bischoff 1986a; Bischoff 1990]. Although IGF-I and FGFs are reported to activate satellite cells, the studies involved typically used cultures of muscle cells that were not quiescent; IGF-I and FGFs increase the proliferative activity of satellite cells once they are activated, even when that activation results during the cell isolation process, i.e. prior to the plating of cells or fibers for culture. Moreover, platelet-derived growth factor BB, transforming growth factor-β (TGF-β), and epidermal growth factor do not stimulate quiescent cells to enter the cell cycle *in vitro* [Bischoff 1986b; Johnson and Allen 1995]. Therefore, HGF is the only growth factor that has been established to have the ability to stimulate quiescent satellite cells to enter the cell cycle early in a culture assay and *in vivo* [Allen et al., 1997; Tatsumi et al., 1998]. HGF is localized to the extracellular domain of uninjured skeletal muscle fibers through a possible association with glycosaminoglycan chains of proteoglycans that are essential components of the extracellular matrix, and following injury, quickly associates with satellite cells [Anderson 2000] by binding to its receptor, c-Met [Tatsumi et al., 1998].

The second component shown to be involved in satellite cell activation is NO, possibly through activation of matrix metalloproteinases, which induce the release of HGF, from the extracellular matrix [Anderson 2000; Tatsumi et al., 2006]. Studies *in vitro* and *in vivo* using rodent muscle have shown HGF and NO to regulate the activity of many satellite cells [Anderson 2000; Miller et al., 2000; Tatsumi et al., 1998; Yamada et al., 2010]. Intriguingly, inhibition of NO production inhibits HGF release, c-Met/HGF co-localization, and satellite cell activation [Anderson 2000]. NO is a short-lived free radical that is well known as a freely diffusible and ubiquitous molecule produced by NOSs from the L-arginine of substrates. In skeletal muscle, neuronal NOS (nNOS, also called NOS-1) is localized to the sarcolemma of muscle fibers by association at its amino terminus with alpha1-syntrophin linked to the dystrophin cytoskeleton [Brenman et al., 1995]. The NO radical is normally produced in very low level pulses by muscles under conditions where satellite cells are quiescent [Tidball et al., 1998], and the expression and activity of constitutive NOS (nNOS and eNOS) are upregulated by exercise, loading injury, shear force, and mechanical stretch. NO also induces expression of follistatin [Pisconti et al., 2006], a fusigenic secreted molecule, known to antagonize myostatin, thus possibly contributing to the exit of satellite cells from quiescence.

More recently, Tatsumi and Allen [Yamada et al., 2010] proposed the intriguing hypothesis that HGF has another role in satellite cells. Although, in culture, a low level of HGF (2.5 ng/ml) optimally stimulates the activation of satellite cells, high levels of HGF (10-500 ng/ml) promote the re-entering of quiescence through a concentration-dependent negative feedback mechanism. Such a role seems to be regulated by the induction of the cyclin-dependent kinase (CDK) inhibitor p21 in a myostatin-dependent manner. Further descriptive analysis is needed to elucidate whether HGF and myostatin really do interact in skeletal muscle *in vivo*. Tatsumi and Allen [Yamada et al., 2010] suggested the importance and difficulty of monitoring whether or not extracellular HGF concentrations reach a threshold (over 10 ng/ml) in muscle of living animals.

# 4. The Proliferating Process of Satellite Cells

## 4.1. Leukemia Inhibitory Factor

Leukemia inhibitory factor (LIF) is a newly discovered myokine [Broholm et al., 2008], originally identified by its ability to induce the terminal differentiation of myeloid leukemic cells. Today, LIF is known to have a wide array of functions, including acting as a stimulus for platelet formation, the proliferation of hematopoietic cells, bone formation, neural survival and formation, muscle satellite cell proliferation and acute phase production by hepatocytes [Metcalf 2003]. LIF is a long chain four α-helix bundle cytokine, which is highly glycosylated and may be present with a weight of 38-67 kDa, which can be deglycosylated to ~20 kDa [Hinds et al., 1997; Schmelzer et al.,

1990]. Several tissues, including skeletal muscle, express LIF. LIF is constitutively expressed at a low level in type I muscle fibers [Kami and Semba 1998; Sakuma et al., 2000] and is implicated in conditions affecting skeletal muscle growth and regeneration [Gregorevic et al., 2002; Kami and Semba 1998; Sakuma et al., 2000]. Indeed, LIF knockout mice showed a decrease in the area occupied by regenerating myofibers after crush injury compared to wild-type mice, which was restored by administration of exogenous LIF [Kurek et al., 1997]. Administration of LIF to the site of crush injury in wild-type mice increased the area occupied by regenerating fibers with an associated increase in average myofiber diameter [Barnard et al., 1994; Kurek et al., 1997]. These original studies suggested that enhanced regeneration and increases in fiber size occurred, at least in part via stimulation of the proliferation of muscle-forming myoblast cells, thus providing more cells to fuse to and increase the size of regenerating fibers.

In 1991, Austin and co-workers demonstrated that LIF stimulated myoblast proliferation in culture [Austin and Burgss 1991], thereby showing that LIF functions as a mitogenic growth factor when added to muscle precursor cells *in vitro*. To date, different groups have confirmed this finding and shown that LIF induces satellite cell and myoblast proliferation, while preventing premature differentiation, by activating a signaling cascade involving Janus kinase 1 (JAK1), signal transducer and activator of transcription (STAT) 1, and STAT3 [Diao et al., 2009; Sun et al., 2007]. In line with this, the specific LIF receptor is primarily expressed by satellite cells and not by mature muscle fibers [Kami et al., 2000]. Thus, it seems that LIF has the potential to affect satellite cells rather than mature muscle fibers.

Earliest descriptions of LIF as a possible mitogen for myoblasts suggested that LIF treatment increased the number of human and mouse-derived primary myoblast cells in a dose-dependent manner after several days of culture, with the earliest increases noticeable after 6 days [Austin and Burgss 1991; Austin et al., 1992]. There is evidence to suggest that LIF promotes survival of myoblasts and other cell types [Negoro et al., 2001; White et al., 2001]. Hunt et al. [2010] found that LIF treatment significantly reduced staurosporine-induced apoptotic DNA fragmentation by 37% and also reduced the proteolytic activation of caspase-3 by 40% compared to controls. This apoptosis-inhibiting role of LIF was completely abolished by a PI3-K (phosphatidylinositol 3-kinase) inhibitor (wortmannin). Therefore, LIF appears to increase the number of satellite cells by promoting proliferation and blocking apoptosis.

## 4.2. Insulin-Like Growth Factor-I and MAPK (Proliferation Phase)

The anabolic effects of IGF-I have been demonstrated in both muscle cell lines and animal models [Adams and McCue 1998; Chakravarthy et al., 2000; Rommel et al., 2001]. For example, the addition of IGF-I to cultured myotubes results in an enlargement of myotube diameters and a higher protein content, while the delivery of IGF-I either through osmotic pumps or genetic overexpression results in increased muscular mass in rodents [Adams and McCue 1998; Musaro et al., 2001]. Mechanical loading also results in skeletal muscle synthesis of IGF-I [Devol et al., 1990; Sakuma et al., 1998] *in vivo*, which stimulates gene expression, DNA and protein synthesis, different transport mechanisms, migration, proliferation, and differentiation [Philippou et al., 2007]. Therefore, investigators conclude that IGF-I is a critical factor involved in skeletal muscle hypertrophy *in vivo* as well as in cultured myotube enlargement *in vitro*.

IGF-I is thought to induce muscle growth through the increased proliferation of satellite cells and the enhancement of protein translation resulting in an increase in the rate of protein synthesis [Clemmons 2009; Philippou et al., 2007]. In addition to stimulating myoblast proliferation, IGF-I stimulates myoblast differentiation [Adi et al., 2002]. For example, IGF-I inhibits production of myogenin, a protein that stimulates muscle cell differentiation, thus allowing increased myoblast proliferation. It is known that the binding of IGF-I to its receptor, after tyrosine (auto)phosphorylation of the receptor, results in the initiation of intracellular cascades of various kinase systems. However, the interplay between the elements of these intracellular signaling pathways has been described based on results of experiments with skeletal muscle cell types of different species and under various conditions. Namely, in mouse and rat skeletal muscle preparations, the involvement of both the MAPK (mitogen-activated protein kinase) pathway and MAPK-independent signaling mechanisms, including PI3-K/Akt and protein kinase C (PKC), was equally documented [Haq et al., 2003; Milasincic et al., 1996; Tiffin et al., 2004]. In primary cultured human skeletal muscle cells, Czifra et al. [2006] demonstrated that the proliferation-enhancing effect of IGF-I was completely inhibited by the PKCδ-specific inhibitor Rottlerin but not by inhibitors of the "conventional" PKCα and γ isoforms or by inhibitors of the MAPK or PI3-K pathway. In addition, overexpression of a kinase inactive mutant of PKCδ prevented the proliferating action of IGF-I. Furthermore, they showed, in mouse C2C12 cells, that the MAPK inhibitor PD098059 partially inhibited the action of IGF-I. Taken together, these results demonstrate a novel, central and exclusive involvement of PKCδ in mediating the action of IGF-I in human skeletal muscle cells, with an additional yet PKCδ-dependent contribution of the MAPK pathway in C2C12 myoblasts.

### 4.3. Notch-Dependent Signaling

The proliferating process in satellite cells appears to be controlled by Notch signaling during muscle regeneration [Conboy and Rando 2002]. Within hours to days following muscle injury, there is increased expression of Notch signaling components (Delta-1, Notch1 and active Notch) in activated satellite cells and neighboring muscle fibers [Conboy and Rando 2002; Conboy et al., 2003]. Upregulation of Notch signaling promotes the transition from activated satellite cells to highly proliferative myogenic precursor cells and myoblasts, as well as prevents differentiation to form myotubes [Buas et al., 2009; Conboy et al., 2003; Kitzmann et al., 2006]. Proliferation was decreased and differentiation was promoted when Notch activity was inhibited in myoblasts with a Notch antagonist, Numb, a gamma-secretase inhibitor, or with small-interfering RNA knockdown of presenilin-1 [Conboy and Rando 2002; Kitzmann et al., 2006; Ono et al., 2009]. In addition, mutations in Delta-like 1 or CSL result in excessive premature muscle differentiation and defective muscle growth [Vasyutina et al., 2007]. Apparent impairment of Notch signaling occurs in aged muscle, because expression of the Notch ligand, Delta, is not upregulated following injury in this muscle. Forced activation of this pathway with a Notch-activating antibody can restore the regenerative potential by inducing the expression of several positive regulators (proliferating cell nuclear antigen, Cyclin D1) of cell cycle progression [Conboy et al., 2003; Conboy et al., 2005].

A recent study revealed that levels of TGF-β are higher in aged than young satellite cell niches [Carlson et al., 2008]. Further analysis showed greater activation of the TGF-β pathway in old satellite cells, and physical competition between Notch and pSmad3 at the promoters of multiple CDK inhibitors [Carlson et al., 2008; Carlson et al., 2009]. Furthermore, the decline of Notch1 signaling with age is thought to be another cause of the decreased regenerative potential of aged skeletal muscle. Indeed, enhancement of Notch1 signaling promotes muscle regeneration in old skeletal muscle [Conboy et al., 2003; Conboy et al., 2005]. Although these experiments suggest a crucial role for Notch1 signaling in satellite cell function, much remains to be determined, especially regarding the role of Notch3 signaling during muscle regeneration. Notch3 was expressed in satellite cells, and various structural and functional differences between Notch3 and Notch1/Notch2 have been reported [Bellavia et al., 2008]. More recently, Kitamoto and Hanaoka [2010] conducted two very intriguing experiments. They analyzed muscle after repeated injuries, by generating mice deficit in Notch3 and also by repetitive intramuscular injections of cardiotoxin (CTX) into the Notch3-deficient mice. They found a remarkable overgrowth of muscle mass in the Notch3-deficient mice but only when they suffered repetitive muscle injuries. Analysis of cultured myofibers revealed that the number of self-renewing Pax7-positive satellite cells attached to myofibers was increased in the Notch3-deficient mice compared to control mice. Given these findings, the Notch3 pathway might act as a Notch1 repressor by activating Nrarp, a negative feedback regulator of Notch signaling.

## 5. The Differentiation of Satellite Cells

### 5.1. MyoD Family

Satellite cell myogenic potential mostly relies on the expression of Pax genes and myogenic regulatory factors (MRFs: MyoD, Myf5, myogenin, and MRF4). Sequential activation and expression of Pax3/7 and MRFs is required for the progression of skeletal myoblasts through myogenesis. Pax7 is expressed by all satellite cells and essential to their postnatal maintenance and self-renewal [Kuang et al., 2006]. Pax7 induces myoblast proliferation and delays their differentiation not by blocking myogenin expression [Zammit et al., 2006] but by regulating MyoD [Olguin et al., 2007]. In parallel, myogenin directly down-regulates Pax7 protein expression during differentiation [Olguin et al., 2007]. MyoD is required for the differentiation of skeletal myoblasts [Cornelison et al., 2000; Sabourin et al., 1999]. In addition, MyoD null satellite cells showed reduced myogenin expression and absolutely no MRF4 expression, and displayed a dramatic differentiation deficit [Cornelison et al., 2000]. Indeed, muscle regeneration *in vivo* is markedly impaired in MyoD null mice [Megeney et al., 1996]. In contrast, Myf5 regulates the proliferation rate and homeostasis [Gayraud-Morel et al., 2007]. MyoD can compensate for Myf5 in adults. Myf5 deficiency leading to a lack of myoblast amplification and loss of MyoD induced an increased propensity for self-renewal rather than progression through myogenic differentiation. The differentiation factors myogenin and MRF4 are not involved in satellite cell development or maintenance [Gayraud-Morel et al., 2007] but induction of myogenin is necessary and sufficient for the formation of myotubes and fibers.

## 5.2. IGF-I and Calcineurin-Dependent Signaling

IGF-I positively regulated not only the proliferation but also the differentiation of satellite cells/myoblasts *in vitro* possibly through a calcineurin-dependent pathway. Since activated calcineurin promotes the transcription and activation of myocyte enhance factor 2 (MEF2), myogenin, and MyoD [Delling et al., 2000; Friday et al., 2000; Friday et al., 2003], calcineurin seems to control satellite cell differentiation and myofiber growth and maturation, all of which are involved in muscle regeneration [Sakuma et al., 2003a; Sakuma et al., 2005]. In fact, our previous study [Sakuma et al., 2003a] showed a marked increase in the amount of calcineurin protein and the clear colocalization of calcineurin and MyoD or myogenin in many myoblasts and myotubes during muscle regeneration. In addition, we showed that the inhibition of calcineurin by cyclosporine A (CsA) induced extensive inflammation, marked fiber atrophy, and the appearance of immature myotubes in regenerating muscle compared with placebo-treated mice [Sakuma et al., 2003a]. Several other studies indicated such defects in skeletal muscle regeneration when calcineurin was inhibited [Abbott et al., 1998; Koulmann et al., 2006], whereas transgenic activation of calcineurin is known to markedly promote the remodeling of muscle fibers after damage [Lara-Pezzi et al., 2007; Stupka et al., 2007].

Many researchers have utilized CsA, though in different amounts, to determine the downstream modulators of calcineurin signaling. We found that intraperitoneal CsA treatment daily at 25 mg/Kg/day enhanced the expression of myostatin and Smad3 mRNA in regeneration-defective tibialis anterior muscle after an injection of bupivacaine [Sakuma et al., 2005]. The possibility that myostatin is a downstream mediator of calcineurin signaling has been indicated by experiments with two different transgenic mice [Michel et al., 2007].

In addition, calcineurin's pharmacological inhibition caused a decline in the transcription and activation of myogenin and MyoD during myogenic differentiation by a downregulation of MyoD expression [Allen and Unterman 2007].

Considering these findings, calcineurin seems to block the myostatin-Smad3 pathway to enhance the expression of myogenic differentiation factor (MyoD) during muscle regeneration *in vivo*. Using CsA treatment *in vivo*, recent evidence including that obtained by our group has also identified Id1 [Friday et al., 2003; Sakuma et al., 2005], Id3 [Friday et al., 2003], and Egr-1 [Friday et al., 2003] as a possible downstream negative hypertrophic effector target of the calcineurin-NFAT (nuclear factor of activated T-cells) pathway.

FOXO (forkhead box O)-induced expression of Atrogin-1 has been shown to inhibit calcineurin activity [Ni et al., 2007]. More recently, the calcineurin variant CnAβ1 was suggested to block the nuclear localization of the FOXO protein and the expression of several genes targeted by FOXO [the muscle ring finger-1 (MuRF1), Gadd45a, Pmaip1, and atrogin genes] in C2C12 myoblasts [Lara-Pezzi et al., 2007]. In addition, transgenic upregulation of CnAβ1 expression promotes the remodeling of CTX-treated muscle fibers [Lara-Pezzi et al., 2007]. In cardiomyocytes, calcineurin directly binds and dephosphorylates (inactivates) Akt; FOXO indirectly activates Akt by inhibiting calcineurin phosphatase activity [Ni et al., 2007]. In murine C2C12 myotubes, Akt was shown to antagonize calcineurin signaling by causing hyperphosphorylation of NFATc1 [Rommel et al., 2001]. Interaction between CnAβ1 and FOXO during muscle regeneration is a very attractive idea, although it has not been demonstrated in adult skeletal muscle *in vivo*.

A more recent study demonstrated the existence of a calcineurin-interacting protein, myospryn [Kielbasa et al., 2011]. Myospryn is a large scaffolding protein localized to the Z-disc/costamere region of striated muscle [Sarparanta 2008]. A defining structural feature of myospryn is a noncanonical tripartite motif (TRIM-like) that lacks the RING domain but consists of a B-box coiled coil, fibronectin 3 repeats, and SPRY domains, which collectively function as a protein-protein interaction interface [Reymond et al., 2001; Sardiello et al., 2008]. Immunoprecipitation by Kielbasa et al. [2011] showed direct binding between calcineurin and myospryn using whole cell-lysates from COS cells (*in vitro*) and neonatal rat ventricular myocytes (*in vivo*). Kielbasa et al. [2011] demonstrated that myospryn inhibits calcineurin-dependent transcriptional activity in C2C12 myoblasts through direct interaction with the enzyme via its TRIM-like domain. In addition, transgenic mice overexpressing the TRIM-like domain of myospryn displayed markedly lower levels of MyoD and myogenin mRNA expression and attenuated muscle regeneration after CTX-induced muscle injury.

### 5.3. Serum Response Factor and MRTF-A

SRF is an ubiquitously expressed member of the MADS box transcription factor family, sharing a highly conserved DNA-binding/dimerization domain, which binds the core sequence of SRE/CArG boxes [CC (A/T)6 GG] as homodimers [Treisman 1987]. Functional CArG boxes have been found in the cis-regulatory regions of various muscle-specific genes, such as the skeletal α-actin [Muscat et al., 1988], muscle creatine kinase, dystrophin, tropomyosin, and myosin light chain 1/3 genes. The majority of SRF's targets are genes involved in cell growth, migration, cytoskeletal organization, and myogenesis [Pipes et al., 2006; Sakuma et al., 2011b].

It is proposed that the transcriptional activity of SRF is regulated by MuRF-2 [Lange et al., 2005] and striated muscle activators of Rho signaling (STARS) [Kuwahara et al., 2005]. At the M-band, the mechanically modulated kinase domain of titin interacts with a complex of the protein products of the atrogenes NBR1, p62/SQSTM-1 and MuRFs [Lange et al., 2005; Puchner et al., 2008]. This complex dissociates under mechanical arrest, and MuRF-1 and MuRF-2 translocate to the cytoplasm and the nucleus [Lange et al., 2005; Ochala et al., 2011]. One of the probable nuclear targets of MuRFs is SRF [Lange et al., 2005], suggesting that the MuRF-induced nuclear export and translocational repression of SRF may contribute to amplifying the transcriptional atrophy program [Spencer et al., 2000]. Thus, it is possible that the synergistic transactivation of SRF and SRF-linked molecules is abrogated by MuRF-2 *in vivo*. On the other hand, SRF activity is extremely sensitive to the state of actin polymerization. G-actin monomers inhibit SRF activity, whereas polymerization of actin occurs in response to serum stimulation and RhoA signaling. In this pathway, signal inputs lower the ratio of globular actin to fibrillar actin thereby liberating the binding of myocardin-related transcription factor (MRTF)-A to globular actin resulting in the nuclear accumulation of MRTF-A, and the subsequent SRF-dependent gene expression of STARS contributes to the nuclear accumulation of MRTF-A and MRTF-B [Kuwahara et al., 2005; Kuwahara et al., 2007]. These factors then activate the translocation of SRF.

SRF was first shown to be essential for both skeletal muscle cell growth and differentiation in experiments performed with C2C12 myogenic cells. In this model, SRF inactivation abolished MyoD and myogenin expression, preventing cell fusion in differentiated myotubes [Soulez et al., 1996]. SRF also enhances the hypertrophic process in muscle fibers after mechanical overloading [Sakuma et al., 2003b]. For example, we showed that, in mechanically overloaded muscles of rats, SRF protein co-localized with MyoD and myogenin in myoblast-like cells during the active differentiation phase [Gauthier-Rouviére et al., 1996]. Interestingly, HAS-Cre:Sf/Sf mice exhibited defects in the regeneration of skeletal muscle after the injection of CTX, although the exact mechanism involved has not been elucidated. The SRF-depleted mice showed decreased levels of IGF-I and IL-4 mRNA at 2 months of age. Since mice with a downregulated IL-4 pathway regenerated normally [Horsley et al., 2003], Charvet et al. [2006] proposed that the regenerative defect was attributable to the decreased expression of IGF-I. During muscle regeneration, the defect of IGF-I expression may affect downstream molecules such as calcineurin [Semsarian et al., 1999] and Akt [Tureckova et al., 2001], although Charvet et al. [2006] did not investigate whether HAS-Cre:Sf/Sf mice had defective calcineurin- and/or Akt-signaling in these skeletal muscles. However, several downstream candidates for these signaling molecules, NFATc2 [Horsley et al., 2001], MyoD [Friday et al., 2000], myogenin [Friday et al., 2003], and myostatin [Michel et al., 2007; Sakuma et al., 2003a], were affected by the SRF mutation [Charvet et al., 2006]. Therefore, the impaired calcineurin- and/or Akt-dependent signaling elicited by the reduction in IGF-I may regulate the regenerative defect recognized in HAS-Cre:Sf/Sf mouse muscles. In contrast, SRF seems not to be required for IGF-I/Akt-dependent muscle growth caused by mechanical overloading [Guerci et al., 2012]. Deletion of Srf from myofibers and not satellite cells blunts overload-induced hypertrophy, and impairs satellite cell proliferation and recruitment to pre-existing fibers. In Srf-deleted muscles, *in vivo* overexpression of COX-2/IL-4 but not IL-6 rescued satellite cell recruitment and muscle growth by enhancing the fusion of satellite cells without affecting their proliferation. These findings support an intriguing hypothesis of Guerci et al. [2012], that SRF translates mechanical cues applied to myofibers into paracrine signals, leading to satellite cell-mediated muscle hypertrophy. However, it is unknown whether the functional role of SRF during muscle hypertrophy applies to the muscle regenerating process because of several important findings demonstrating the outstanding differences in the role of satellite cells in these adaptations [McCarthy et al., 2011].

More recently, Mokalled et al. [2012] demonstrated that members of the Myocardin family of transcriptional coactivators, MASTR and MRTF-A, are upregulated in satellite cells in response to skeletal muscle injury. In addition, global and satellite cell-specific deletions of MASTR in mice impair skeletal muscle regeneration.

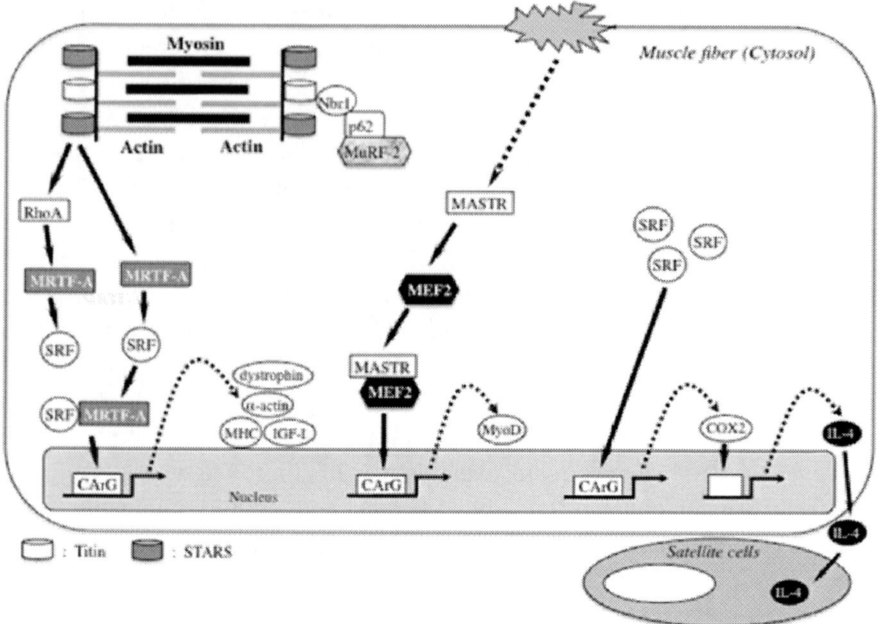


Figure 1. Schematic diagram of SRF-dependent signaling in muscle regeneration. Mechanical loading produced by muscle contraction causes myosin and actin to interact, which in turn activates STARS and titin. STARS protein activates MRTF-A indirectly via RhoA or directly [Kuwahara et al., 2005]. Activated MRTF-A binds to SRF to promote the expression of muscle-specific genes such as those for α-actin, dystrophin, IGF-I, and myosin heavy chain (MHC). The zinc finger protein Nbr1 binds to both titin and p62 at the N-terminal PB1 domain [Lange et al., 2005]. In normal muscle, p62 binds to the ubiquitin ligase MuRF-2, via an ubiquitin-associated domain at its C-terminus. In the differentiation phase of muscle regeneration, activated MASTR binds to MEF2 to upregulate the expression of MyoD [Mokalled et al., 2012]. In mechanically overloaded muscle (probably also in the regenerating muscle), SRF enhances the expression of COX2 mRNA, which in turn upregulates IL-4 mRNA, and ultimately secretes IL-4 protein [Guerci et al., 2012]. IL-4 produced by muscle fibers moves into satellite cells paracrinally to modulate the fusion of satellite cells. COX2; cyclooxygenase 2, IGF-I; insulin-like growth factor-I, IL-4; interleukin-4, MEF2; myocyte enhancer hactor 2, MHC; myosin heavy chain, MRTF-A; myocardin-related transcription factor-A, SRF; serum response factor, STARS; striated muscle activators of Rho signaling.

This impairment is substantially greater when MRTF-A is also deleted and is due to aberrant differentiation and excessive proliferation of satellite cells. In fact, double knockout satellite cells (MASTR and MRTF-A) reveal a significant downregulation of the various modulators of cell cycle arrest such as CyclinG1, Retinoblastoma, growth-arrest specific 2, and growth arrest and DNA damage-inducible 45a. In myogenesis, MASTR activates a muscle-specific postnatal MyoD enhnacer through associations with MEF2 and members of the myocardin family. It remains to be elucidated whether this MASTR functions by directly binding with SRF. Using the Xenopus, Meadows et al. [2008] showed that MASTR promoted the expression of skeletal muscle-specific genes by co-operating with SRF. The muscle-marker's activation was dependent on the ability to interact with SRF, because a mutant form of MASTR lacking the SRF-binding domain failed to induce such expression [Meadows et al., 2008]. However, the mouse MASTR protein lacks SRF-interacting regions, and so its assembly into a transcription regulatory complex would rely on interactions with other factors (e.g., MEF2 proteins). Mokalled et al. [2012] proposed the intriguing hypothesis that interaction between MASTR and MEF2 (MEF2A and MEF2C) enhances the expression of cell cycle arrest genes and MyoD expression at the transition from proliferation to differentiation of satellite cells. Figure 1 summarizes the functional role of SRF and SRF-linked molecule (MASTR) in muscle regeneration.

## 6. Second-Stage Fusion (Late Differentiation)

A central pathway involved in hypertrophy is regulated at the translational level by the serine/threonine kinase Akt. In muscle, Akt is activated by the upstream PI3-K, induced either by receptor binding or by integrin-mediated activation of focal adhesion kinase, such as in cardiac myocytes [Franchini et al., 2000; Sakamoto et al., 2002]. The striking effect of Akt1 on muscle size was demonstrated by the transient transfection of a constitutively active inducible Akt1 transgene in skeletal muscle *in vivo* [Lai et al., 2004]. In addition, muscle mass was completely preserved in denervated transgenic Akt mice [Sartori et al., 2009]. Possible downstream regulators of Akt, mammalian target of rapamycin (mTOR) and glycogen synthase 3-β play a crucial role in the regulation of translation [Cross et al., 1995].

More recently, it has been shown that mTOR exists in two functionally distinct multi-protein signaling complexes, mTOR signaling complex (mTORC)1 and mTORC2 [Jacinto et al., 2004]. Akt activates mTOR via phosphorylation and inactivation of tuberous sclerosis complex-2 [Manning et al., 2002]. In general, only signaling by mTORC1 is inhibited by rapamycin, and thus the growth regulatory effects of rapamycin are believed to be primarily exerted through the mTORC1 complex [Zoncu et al., 2011]. It is now widely accepted that signaling by mTORC1 is involved in the regulation of several anabolic processes including protein synthesis, and ribosome biogenesis as well as catabolic processes such as autophagy [Zoncu et al., 2011]. In skeletal muscle, signaling by mTORC1 is activated in response to hypertrophic stimuli such as increased mechanical loading, feeding and growth factors [Drummond et al., 2009; Rommel et al., 2001].

Growth and maturation of the muscle cells are achieved through a second-stage fusion, which occurs between the nascent myofibers/myotubes and myoblasts. Although many regulators of this fusion process have been revealed in recent years [Jansen and Pavlath 2008], a better understanding of the regulation is still needed. mTOR is one of the candidates regulating the fusion. mTOR signaling regulates a wide range of biological processes, including cell growth, various types of cellular differentiation, and metabolism [Sarbassov et al., 2005; Wullschleger et al., 2006]. mTOR assembles two biochemically and functionally distinct protein complexes, mTORC1 and mTORC2, which are sensitive and insensitive to rapamycin, respectively [Sarbassov et al., 2005]. Rapamycin-sensitive mTORC1 signaling has emerged as a key regulator of skeletal muscle differentiation and remodeling. Rapamycin inhibits myoblast differentiation *in vitro* [Cuenda and Cohen 1999; Erbay and Chen 2001], compensatory myofiber hypertrophy *in vivo*, and regrowth of myofibers after atrophy [Bodine et al., 2001]. The regulation of skeletal myocyte differentiation by mTORC1 occurs at two stages via distinct mechanisms. mTORC1 controls the initiation of myoblast differentiation by regulating IGF-II expression [Erbay and Chen 2001], whereas late-stage myocyte fusion leading to myotube maturation is regulated by mTORC1 through a yet to be identified secreted factor [Park and Chen 2005]. More recent findings pointed out that the fusion factor targeting mTORC1 is follistatin during the late differentiation phase. Sun et al. [2010] have found that, in C2C12 cells differentiating for 24-72h, miR-1 luciferase (enhancer) activity was markedly downregulated after treatment with rapamycin but not wortmannin (PI3-K inhibitor) or SB203580 (MAPK inhibitor). In addition, rapamycin increased the amount of histone deacetylase (HDAC)4 protein and reduced follistatin mRNA and MyoD protein levels in C2C12 and C3H10T1/2 cells. Furthermore, daily administration of tricostatin A and a single dose of adenovirus expressing follistatin rescued the defective muscle regeneration caused by treatment with rapamycin. Sun et al. [2010] proposed the intriguing hypothesis that mTOR-miR-1 promotes myocyte fusion by recruiting HDAC4-

follistatin during myoblast differentiation *in vitro* and skeletal muscle regeneration *in vivo.*

## 7. Satellite Cell Self-Renewal

A hallmark of stem cells is their ability to self-renew. In skeletal muscle, asymmetric cell division takes place in a subset of the satellite cell population to generate a self-renewing progenitor and hyperplastic daughter cell that later contributes to *de novo* muscle formation [Kuang et al., 2007]. Several extrinsic pathways have been implicated in mediating this phenomenon [Kuang et al., 2007; Le Grand et al., 2009]. One family of candidate peptides is the Wnt family of signaling molecules which consists of over 19 cysteine-rich secreted glycoproteins that in part bind the Frizzled receptors [Van Amerongen and Nusse 2009].

In a non-canonical Wnt cascade, Wnt7a has been characterized for its role as the extracellular ligand mediating asymmetric cell division that is thought to be the mechanism by which satellite cells are able to self-renew [Le Grand et al., 2009]. Lineage tracing of satellite cell populations indicates ~90% of cells to have at some point expressed Myf5 ($Pax7^{+}Myf5^{+}$) [Kuang et al., 2007]. The $Myf5^{+}$ cells have a reduced potential to self-renew as the majority undergo symmetrical cell divisions and later contribute to muscle syncitia [Kuang et al., 2007]. The remaining ~10% of satellite cells divide asymmetrically and give rise to $Pax7^{+}Myf5^{-}$ as well as $Pax7^{+}Myf5^{+}$ progeny thereby maintaining the stem cell pool of muscle progenitors [Kuang et al., 2007]. The capacity of $Pax7^{+}Myf5^{-}$ cells to self-renew is explained by expression of the Wnt receptor Fzd7 on these cells but not on $Pax7^{+}Myf5^{+}$ cells, thus allowing induction of asymmetrical cell division via Wnt7a-induced signaling [Le Grand et al., 2009]. Importantly, stimulation of satellite cells with Wnt7a leads to an increase in the symmetrical expansion of satellite cells, while muscle from Wnt7a knockout mice displays a dramatic reduction in satellite cell numbers following regeneration [Le Grand et al., 2009].

The Notch inhibitor, Numb is also asymmetrically expressed on the activated satellite cells and may regulate cell fate choices by promoting progression down the myogenic lineage [Conboy and Rando 2002]. Self-renewal may also occur through symmetrical division in which both daughter cells maintain stem-cell properties [Cosgrove et al., 2009; Morrison and Kimble 2006]. Cells that do not express MyoD but continue to express Pax7 are suggested to be refrained from self-renewal [Tajbakhsh 2009].

# 8. Other Regulators of the Muscle Regenerating Process

## 8.1. Myostatin and TGF-β

The transforming growth factor-beta (TGF-β) superfamily plays a crucial role in normal physiology and pathogenesis in a number of tissues. Myostatin was first discovered during screening for novel members of the TGF-β superfamily, and shown to be a potent negative regulator of muscle growth [Lee 2004]. Like other TGF-β family members, myostatin is synthesized as a precursor protein that is cleaved by furin proteases to generate the active C-terminal dimer. When produced in Chinese hamster ovary cells, the C-terminal dimer remains bound to the N-terminal propeptide, which remains in a latent, inactive state [Wolfman et al., 2003]. Most, if not all, of the myostatin protein that circulates in blood also appears to exist in an inactive complex with a variety of proteins, including the propeptide [Zimmers et al., 2002]. Myostatin binds to and signals through a combination of Activin IIA/B (ActRIIA/B) receptors on the cell membrane, but has higher affinity for ActRIIB. On binding to ActRIIB, myostatin forms a complex with a second surface type I receptor, either activin receptor-like kinase (ALK4 or ActRIB) or ALK5 to stimulate the phosphorylation of receptor Smad and the Smad2/3 transcription factors in the cytoplasm. This leads to the assembly of Smad2/3 with Smad4 to form a heterodimer that is able to translocate to the nucleus and activate the transcription of target genes [Joulia-Ekaza and Cabello 2007].

Studies indicate that myostatin inhibits the activation, differentiation, and self-renewal of satellite cells [Langley et al., 2002; McCroskery et al., 2003; Yang et al., 2007] and the expression of the muscle regulatory factors crucial for the regeneration and differentiation of myofibers [Joulia et al., 2003; Langley et al., 2002]. Intriguingly, loss of Smad3, a possible mediator for myostatin, also led to defective satellite cell functionality. Indeed, Ge et al., [2011] observed the decreased satellite cell numbers in skeletal muscle from Smad3-null myoblasts probably due to the reduced propensity for self-renewal. Furthermore, *in vitro* analysis of primary myoblast cultures identified that Smad3-null myoblasts exhibit impaired proliferation, differentiation, and fusion. A more recent study [Ge et al., 2012] clearly indicated that the mice with null mutation of Smad3, an intracellular mediator for both myostatin and TGF-β, exhibited incomplete recovery of muscle weight and myofiber size after muscle injury. Morphological analysis suggested impaired inflammatory response and a reduced number of activated myoblasts during the early stage of muscle regeneration in Smad3-null mice. In addition, Smad3-null regenerated muscle had decreased oxidative enzyme activity and impaired mitochondrial biogenesis possibly due to the down-regulation of the gene encoding TFAM, a master regulator of mitochondrial biogenesis.

TGF-β1 is expressed during myogenesis, and its spatial and temporal expression in the developing connective tissue is correlated with the fiber-type composition of the surrounding myotubes. Myotubes formed before the expression of TGF-β1 develop into slow fibers, whereas fast fibers form when myoblasts are adjacent to connective tissue expressing TGF-β1 [McLennan 1993]. TGF-β1 has been shown to inhibit the differentiation of fetal myoblasts but does not affect embryonic myoblasts [Cusella-De Angelis et al., 1994]. In mature adult muscle, TGF-β negatively affects skeletal muscle regeneration by inhibiting satellite cell proliferation, myoblast fusion, and expression of some muscle specific-genes [Allen and Boxhorn 1987]. Furthermore, TGF-β1 induced the transformation of myogenic cells into fibrotic cells after injury [Li et al., 2004].

TGF-β1, a potent regulator of tissue wound healing and fibrosis, is physiologically upregulated in regenerating skeletal muscle after injury and exercise and is thought to participate in a transient inflammatory response to muscle damage [Gosselin and McCormick 2004; Serrano and Munoz-Canoves 2010]. Persistent exposure to the inflammatory response leads to an altered extracellular matrix and increased levels of growth factors and cytokines, including TGF-β1, which contribute to the formation of fibrotic tissue [Gosselin and McCormick 2004; Serrano and Munoz-Canoves 2010]. Increased levels of TGF-β1 inhibit satellite cell activation and impair myocyte differentiation [Allen and Boxhorn 1987; Allen and Boxhorn 1989].

## 8.2. TNF-α Signaling

TNF-α has long been viewed as the quintessential proinflammatory cytokine, capable of classical activation of macrophages to the M1 phenotype, and thereby inducing the production of other proinflammatory, Th1 cytokines. Following muscle injury, the early invading neutrophil and macrophage populations express TNF-α [Zádor et al., 2001], suggesting that the cytokine may contribute to the early inflammatory stages that precede muscle regeneration. TNF-α levels in muscle following acute injury peak at 24h postinjury, which indicates that TNF-α production is most tightly coupled with the Th1 inflammatory response in injured muscle [Warren et al., 2002]. Because findings show that TNF-α induces iNOS expression in myeloid cells and that myeloid cell-derived NO can cause muscle fiber damage early on, Th1 inflammatory cells have been associated with muscle damage. However, TNF-α levels remain elevated for nearly 2 weeks following acute injury, indicating that TNF-α may also modulate the regenerative process [Warren et al., 2002]. Intriguingly, the expression of TNF-α receptors by muscle cells themselves is elevated as a later consequence of injury, during the regenerative process, and enables TNF-α to act directly on muscle cells to modulate their proliferation and differentiation [Zádor et al., 2001].

Numerous experimental observations indicate that TNF-α acts directly on muscle cells in affecting muscle regeneration. For example, TNF-α null mutants and TNF-α receptor mutants show lower levels of MyoD and MEF2 expression than wild-type controls following acute injury [Chen et al., 2005; Warren et al., 2002]. The application of exogenous TNF-α to myoblasts *in vitro* increases their proliferation, and inhibited the process of early differentiation to terminal differentiation [Guttridge et al., 1999; Langen et al., 2001; Langen et al., 2004]. Experiments *in vivo* using a lung-specific TNF-α transgene also showed a differentiation-inhibiting role [Langen et al., 2006]. These TNF-α abundant mice exhibited attenuated expression of developmental MHC in reloaded soleus muscle after hindlimb suspension [Langen et al., 2006]. TNF-α affects several intracellular signaling pathways leading to the activation of NF-κB, caspase 8, and stress-induced factors like c-Jun N-terminal kinase (JNK) and p38 MAPK [Guttridge 2004]. Activation of NF-κB can inhibit myogenesis through several processes. NF-κB can promote the expression and stability of cyclin D1 in muscle [Guttridge et al., 1999], leading to increased cell proliferation and inhibition of differentiation. Furthermore, NF-κB can cause destabilization of MyoD mRNA and degradation of MyoD protein [Guttridge et al., 1999; Langen et al., 2001]. The role of JNK in the effect of TNF-α on myogenesis has been less investigated. A recent study suggested that activation of JNK by TNF-α blocks IGF-I signaling necessary for the differentiation of myoblasts [Strle et al., 2006].

TNF-α can activate signaling through other pathways independent of NF-κB to promote muscle differentiation. Both IL-1 and TNF-α can activate p38 kinase [Raingeaud et al., 1995], promoting the differentiation. In particular, inhibition of p38 in skeletal muscle cells *in vitro* inhibits myocytes from fusing to form myotubes and reduces the expression of MEF2, myogenin, and myosin light chain kinase [Zetser et al., 1999], all of which indicate that p38 activation can promote muscle differentiation. Furthermore, p38 activation can also increase the activity of MyoD [Wu et al., 2000; Zetser et al., 1999]. The ability of p38 to promote myogenesis relies, in part, on its ability to phosphorylate and increase the transcriptional activity of MEF2 [Han et al., 1997; Zetser et al., 1999]. In contrast, p38 activation can also inhibit myogenesis by the phosphorylation of other MyoD family members (MRF4). The elevated expression and activity of p38 late in muscle differentiation leads to increased MRF4 phosphorylation and, as a consequence, a decline in desmin and skeletal α-actin expression [Suelves et al., 2004]. In fact, overexpression of MRF4 in a transgenic mouse line caused defective muscle regeneration following injury [Pavlath et al., 2003]. Therefore, TNF-α-dependent signaling regulates various aspects of the muscle regenerating process (immune response, and proliferation and differentiation of satellite cells) through different downstream mediators (NF-κB, JNK, and p38) [Figure 2].

## 8.3. NF-κB Signaling

NF-κB refers to structurally related Rel family eukaryotic transcription factors, which regulate a variety of cellular responses [Acharyya et al., 2007]. The NF-κB family constitutes five members, which can be further divided in two groups. One group includes RelA (p65), RelB, and c-Rel, which are synthesized as mature proteins and characterized by the presence of an N-terminal Rel homology domain essential for dimerization and DNA-binding and a C terminal transcriptional activation domain. The second group consists of the NF-κB1 (p50) and NF-κB2 (p52) proteins, which are synthesized as the larger precursors p105 and p100, respectively, containing an N-terminal ankyrin repeat domain. Proteolytic processing of p105 and p100 at the C terminus gives rise to p50 and p52, respectively. Both p50 and p52 contain the N-terminal Rel homology domain but lack the transcriptional activation domain at the C terminus [Acharyya et al., 2007].

Different members of the NF-κB family dimerize to facilitate the binding of NF-κB to DNA. Among then, p50 and p65 are the most prototypical heterodimers, present in almost all cell types and responsible for the increased expression of a number of pro-inflammatory and cell survival genes. However, homodimers or heterodimers of p50 and p52, which lack transcriptional activation domains, can still bind to NF-κB consensus sites in DNA and act as transcriptional repressors by blocking the consensus sites [Hayden and Ghosh 2008]. Prior to activation, most NF-κB dimers are retained in the cytoplasm by binding to specific inhibitors-the inhibitors of NF-κB (IκBs). The interaction with IκBs masks the nuclear localization sequence in the NF-κB complex, thus preventing nuclear translocation and maintaining NF-κB in an inactive state in the cytoplasmic compartment [Hayden and Ghosh 2008].

### *8.3.1. NF-κB Regulates Both the Proliferation and Differentiation in Myogenesis*

Contrary to the findings above, separate reports have associated NF-κB with a negative regulatory rokle in skeletal muscle differentiation. In several laboratories, NF-κB DNA binding activity was found to decline over the course of myogenesis [Bakkar et al., 2005; Catani et al., 2004; Guttridge et al., 1999]. This regulation was accompanied by a reduction in NF-κB transcriptional activity as recorded from reporter assays and from expression of a bona fide target of NF-κB, IκBα [Guttridge et al., 1999]. Additionally, inhibition of NF-κB signaling by stable expression of an IκBα-super repressor inhibitor mutant was found to accelerate myogenesis [Guttridge et al., 1999]. NF-κB can act at multiple levels to block muscle differentiation. CyclinD1, itself a reported repressor of myogenesis [Skapek et al., 1995] is also a transcriptional target of NF-κB [Guttridge et al., 1999]. The cyclin D1 protein was recently reported to interact and be stabilized by p65 [Dahlman et al., 2009]. In addition, classical NF-κB subunits can suppress the synthesis of MyoD by acting through a destabilization element in the MyoD transcript in response to TNF-α and TWEAK signaling [Dogra et al., 2006; Guttridge et al., 2000]. Furthermore, NF-κB was shown to inhibit myogenesis in proliferating myoblasts through activation of Yin-Yangl (YY1). In muscle cells, YY1 functions as a transcriptional repressor by associating with Ezh2 and the Polycomb group to silence myofibrillar genes that include, but may not be necessarily limited to, the troponin C, MHC IIb, and a-actin genes [Wang et al., 2007].

### *8.3.2. NF-κB Function in Muscle Regeneration*

As a regulator of myogenesis, classical NF-κB has also been found to modulate muscle regeneration both in response to damage and in degenerative muscle diseases. In a model of CTX injury model, a lack of p65 in 4-week-old mice was accompanied by increased numbers of centrally located nuclei, a hallmark of muscle regeneration [Wang et al., 2007]. Similarly, mice lacking the classical kinase IKK (IκB kinase)β specifically in skeletal muscles showed enhanced regeneration as revealed by repaired fibers of increased size [Mourkioti et al., 2006]. Mechanistically, Mourkioti et al. [2006] observed increased numbers of centrally located myonuclei per regenerated fiber in

IKKβ-deleted muscles. Furthermore, these muscles accumulated less fibrotic tissue and exhibited an earlier clearance of inflammatory infiltrates, correlating with enhanced muscle regeneration. Guttridge's laboratory linked the repair process to increased numbers of muscle progenitors, namely, a CD34+/Sca-1− population coinciding with Pax7-positive satellite cells. They also reported that muscle-specific inhibition of IKKβ led to decreased levels of TNF-α, thus implying that mature muscles are capable of producing this cytokine. Given that TNF-α has been found to be a potent inhibitor of skeletal myogenesis when administered at nonphysiological levels [Guttridge et al., 2000; Langen et al., 2002], one can postulate that NF-κB/IKKβ represses regeneration in dystrophic muscles by promoting the secretion from myofibers of TNF-α which then signals to satellite cells or myoblasts to inhibit their differentiation. Taken together, these studies support that disruption of classical NF-κB signaling in mature muscles enhances regenerative myogenesis and conversely that this pathway negatively regulates adult muscle differentiation.

## 8.4. TWEAK

TNF-like weak inducer of apoptosis (TWEAK) is a pro-inflammatory cytokine belonging to the TNF superfamily of ligands. Initially synthesized as a type II transmembrane protein, TWEAK is cleaved to its soluble form, and signals as a trimerized molecule [Winkles 2008]. Generally, TWEAK signaling occurs through binding to fibroblast growth factor-inducible 14 (Fn14), a type I transmembrane receptor first recognized using a differential display technique and later identified as the TWEAK receptor [Meighan-Mantha et al., 1999; Winkless et al., 2007]. Its cytoplasmic domain contains a TNF-receptor-associated factor (TRAF)-binding site that allows recruitment of various TRAFs, which are also involved in cell signaling by other members of TNFSF [Brown et al., 2003]. TWEAK-Fn14 dyad regulates several physiological responses including cell survival, proliferation, angiogenesis, migration, and apoptosis [Winkles 2008].

Dogra et al. [2006] reported that TWEAK inhibits the differentiation of cultured C2C12 or primary myoblasts into multinucleated myotubes. TWEAK has been found to regulate the regeneration and growth of myofibers after injury [Dogra et al., 2006; Dogra et al., 2007a; Dogra et al., 2007b]. The role *in vivo* of TWEAK in skeletal muscle regeneration has now been investigated employing both TWEAK-KO and muscle-specific TWEAK-transgenic (Tg) mice [Mittal et al., 2010]. The expression of both TWEAK and Fn14 increased significantly within 3-5 days of the injection of CTX. When muscle regeneration was evaluated, no obvious difference in muscle structure was observed between wild-type. TWEAK-KO, and TWEAK-Tg mice, 5 days after the CTX injection. However, at 10 and 21 days post-injection, regenerating myofibers of TWEAK-KO mice appeared larger in diameter compared to those of wild-type mice [Mittal et al., 2010]. By contrast, regenerating fibers were smaller in TWEAK-Tg mice than wild-type littermates [Mittal et al., 2010]. Further analysis of muscle using biochemical and histological techniques showed that TWEAK mediates the inflammatory response leading to diminished regeneration and /or growth. In fact, mRNA levels of TNF-α, IL-6 and CCL-2 and protein levels of embryonic MHC were significantly reduced in CTX-injected TA muscle of TWEAK-KO mice compared to that of wild-type mice [Mittal et al., 2010]. In addition, these parameters were found to be significantly increased in regenerating TA muscle of TWEAK-Tg mice compared to that of control mice. Since such a modulation of the TWEAK gene caused no apparent differences in levels of phospho-Akt and phospho p38MAPK in the regenerating muscle among each mouse model, TWEAK seems to function independently of Akt- and p38-linked signaling [Mittal et al., 2010]. Intriguingly, electromobility shift assay [Mittal et al., 2010] indicated the possibility of TWEAK-NF-κB signaling, although further descriptive analysis needs to be done.

It is interesting to note that the role of TWEAK and Fn14 in adult skeletal muscle regeneration is quite similar to their individual roles in myogenic differentiation. Fn14-KO mice showed delayed muscle regeneration after injury [Girgenrath et al., 2006]. The number of newly formed fibers with centronucleation and/or positive for the embryonic form of MHC was significantly reduced in Fn14-KO mice compared to wild-type mice in TA muscle at 5 and 7 days following the injection of CTX [Girgenrath et al., 2006]. This study also suggested the delayed muscle regeneration in Fn14-KO mice to be due to a diminished/delayed inflammatory response. Within 1-3 days post CTX injection, Fn14-KO mice showed lower numbers of macrophages and neutrophil infiltrates in muscle tissues. Figure 2 indicates the possible role of TNF-α and TWEAK in various aspects of the muscle regenerating process (immune response, and proliferation and differentiation of satellite cells).

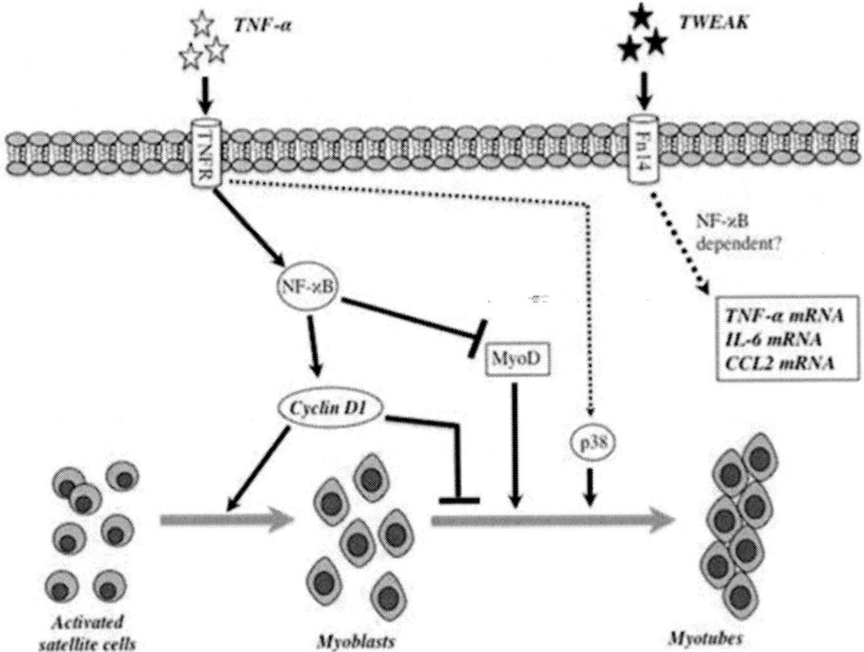


Figure 2. The functional role of TNF-α signaling in the regenerating muscle. In regenerating muscle after treatment with myotoxin, the differentiating myotubes seem to be fused together and/or incorporated into the existing muscle fibers. TNF-α, which is produced by the damaged muscle and macrophages, stimulates TNFR. TNFR activates NF-κB-signaling, in turn cyclin D1 activate the proliferation, but not differentiation, of satellite cells. In contrast, TNF-α/NF-κB signaling blocks the differentiation-promoting role by MyoD. In addition, TNF-α activates p38-dependent signaling leading to the differentiation of myoblasts. The interaction between TWEAK and Fn14 upregulates the gene expression of TNF-α, IL-6, and CCL2 to induce appropriate inflammatory response probably via NF-κB. CCL; chemokine (C-C motif) ligand, Fn14; fibroblast growth factor-inducible 14, IL-6; interleukin-6, NF-κB; nuclear factor-kappaB, TNF-α; tumor necrosis-factor-α, TNFR: TNF-α receptor, TWEAK: TNF-like weak inducer of apoptosis.

## Conclusions and Perspectives

In normal, skeletal muscle possesses a robust capacity to repair itself, the ability to augment and enhance this process would significantly advance the treatment of congenital muscle disorders and severe muscle trauma for which, even with the best of present-day treatments, physical handicap or amputation are the most likely outcomes. Sarcopenia seems to include the defect of muscle regeneration probably due to the repetitive muscular damage [Carlson et al., 2008; Conboy et al., 2003; Conboy et al., 2005]. Currently available data show that resistance training combined with amino acid-containing supplements would be the best way to prevent age-related muscle wasting and weakness. Therfore, for these endogenous repair therapies to advance, it is essential that an understanding exists of the biochemical, cellular and mechanical cues that promote skeletal muscle repair.

## Acknowledgments

This work was supported by a research Grant-in-Aid for Scientific Research C (No. 23500578) from the Ministry of Education, Culture, Sports, Science and Technology of Japan.

## References

Abbott, K. L., Friday, B. B., Thaloor, D., Murphy, T. J. and Pavlath, G. K. (1998) Activation and cellular localization of the cyclosporine A-sensitive transcription factor NF-AT in skeletal muscle cells. *Mol. Biol. Cell*, 9, 2905-2916.

Acharyya, S., Villalta, S. A., Bakkar, N., Bupha-Intr, T., Janssen, P. M., Carathers, M., Li, Z. W., Beg, A. A., Ghosh, S., Sahenk, Z., Weinstein, M., Gardner, K. L., Rafael-Fortney, J. A., Karin, M., Tidball, J. G., Baldwin, A. S. and Guttridge, D. C. (2007) Interplay of IKK/NF-kappaB signaling in macrophages and myofibers promotes muscle degeneration in Duchenne muscular dystrophy. *J. Clin. Invest.*, 117, 889-901.

Adams, G. R. and McCue, S. A. (1998) Localized infusion of IGF-I results in skeletal muscle hypertrophy in rats. *J. Appl. Physiol.*, 84, 1716-1722.

Adi, S., Bin-Abbas, B., Wu, N. Y. and Rosenthal, S. M. (2002) Early stimulation and late inhibition of extracellular signal-regulated kinase 1/2 phosphorylation by IGF-I: A potential mechanism mediating the switch in IGF-I action on skeletal muscle cell differentiation. *Endocrinology*, 143, 511-516.

Allen, R. E. and Boxhorn, L. K. (1987) Inhibition of skeletal muscle satellite cell differentiation by transforming growth factor-beta. *J. Cell. Physiol.*, 133, 567-572.

Allen, R. E. and Boxhorn, L. K. (1989) Regulation of skeletal muscle satellite cell proliferation and differentiation by transforming growth factor-beta, insulin-like growth factor I, and fibroblast growth factor. *J. Cell. Physiol.*, 138, 311-315.

Allen, R. E., Temm-Grove, C. J., Sheehan, S. M., and Rice, G. M. (1997) Skeletal muscle satellite cell cultures. *Methods Cell Biol.*, 52, 155-176.

Allen, D. L. and Unterman, T. G. (2007) Regulation of myostatin expression and myoblast differentiation by FoxO and SMAD transcription factors. *Am. J. Physiol. Cell Physiol.*, 292, C188-C199.

Al-Shanti, N., Saini, A., Faulkner, S. H. and Stewart, C. E. (2008) Beneficial synergistic interactions of TNF-alpha and IL-6 in C2 skeletal myoblasts-potential cross-talk with IGF system. *Growth Factors*, 26, 61-73.

Al-Shanti, N. and Stewart, C. E. (2009) $Ca^{2+}$/calmodulin-dependent transcriptional pathways: Potential mediators of skeletal muscle growth and development. *Biol. Rev. Camb. Philoso. Soc.*, 84, 637-652.

Anderson, J. E. (2000) A role for nitric oxide in muscle repair: NO-mediated satellite cell activation. *Mol. Biol. Cell*, 11, 1859-1874.

Arnold, L., Henry, A., Poron, F., Baba-Amer, Y., Van Rooijen, N., Plonquet, A., Gherardi, R. K. and Chazaud, B. (2007) Inflammatory monocytes recruited after skeletal muscle injury switch into anti-inflammatory macrophages to support myogenesis. *J. Exp. Med.*, 204, 1057-1069.

Austin, L. and Burgss, A. W. (1991) Stimulation of myoblast proliferation in culture by leukaemia inhibitory factor and other cytokines. *J. Neurol. Sci.,* 101, 193-197.

Austin, L., Bower, J., Kurek, J. and Vakakis, N. (1992) Effects of leukaemia inhibitory factor and other cytokines on murine and human myoblast proliferation. *J. Neurol. Sci.*, 112, 185-191.

Bakkar, N., Wackerhage, H. and Guttridge, D. C. (2005) Myostatin and NF-κB regulate skeletal myogensis through distinct signaling pathways. *Signal Transduction*, 4, 202-210.

Barnard, W., Bower, J., Brown, M. A., Murphy, M. and Austin, L. (1994) Leukemia inhibitory factor (LIF) infusion stimulates skeletal muscle regeneration after injury: Injured muscle expresses LIF mRNA. *J. Neurol. Sci.*, 123, 108-113.

Bellavia, D., Checquolo, S., Campese, A. F., Felli, M. P., Gulino, A. and Screpanti, I. (2008) Notch3: From subtle structural differnces to functional diversity. *Oncogene,* 27, 5092- 5098.

Bischoff, R. (1986a) A satellite cell mitogen from crushed adult muscle. *Dev. Biol.,* 115, 140-147.

Bischoff, R. (1986b) Proliferation of muscle satellite cell on intact myofibers in culture. *Dev. Biol.*, 115, 129-139.

Bischoff, R. (1990) Cell cycle commitment of rat muscle satellite cells. *J. Cell Biol.,* 111, 201-207.

Bodine, S. C., Stitt, T. N., Gonzalez, M., Kline, W. O., Stover, G. L., Bauerlein, R., Zlotchenko, E., Scrimgeour, A., Lawrence, J. C., Glass, D. J. and Yancopoulos, G. D. (2001) Akt/mTOR pathway is a crucial regulator of skeletal muscle hypertrophy and can prevent muscle atrophy in vivo. *Nat. Cell Biol.*, 3, 1014-1019.

Brenman, J. E., Chao, D. S., Xia, H., Aldape, K. and Bredt, D. S. (1995) Nitric oxide synthase complexed with dystrophin and absent from skeletal muscle sarcolemma in Duchenne muscular dystrophy. *Cell*, 82, 743-752.

Broholm, C., Mortensen, O. H., Nielsen, S., Akerstrom, T., Zankari, A., Dahl, B. and Pedersen, B. K. (2008) Exercise induces expression of leukaemia inhibitory factor in human skeletal muscle. *J. Physiol.*, 586, 2195-2201.

Brown, S. A., Richards, C. M., Hanscom, H. N., Feng, S. L. and Winkles, J. A. (2003) The Fn14 cytoplasmic tail binds tumor-necrosis-factor-receptor-associated factors 1, 2, 3 and 5 and mediates nuclear factor-kappaB activation. *Biochem. J.,* 371, 395-403.

Bryer, S. C., Fantuzzi, G., Van Rooijen, N. and Koh, T. J. (2008) Urokinase-type plasminogen activator plays essential roles in macrophage chemotaxis and skeletal muscle regeneration. *J. Immunol.*, 180, 1179-1188.

Buas, M., Kabak, S. and Kadesch, T. (2009) Inhibition of myogenesis by Notch: Evidence for multiple pathways. *J. Cell. Physiol.*, 218, 84-93.

Carlson, M. E., Hsu, M. and Conboy, I. M. (2008) Imbalance between pSmad3 and Notch induces CDK inhibitors is old muscle stem cells. *Nature,* 454, 528-532.

Carlson, M. E., Conboy, M. J., Hsu, M., Barchas, L., Jeong, J., Agrawal, A., Mikels, A. J., Agrawal, S., Schaffer, D. V. and Conboy, I. M. (2009) Relative roles of TGF-beta1 and Wnt in the systemic regulation and aging of satellite cell responses. *Aging Cell*, 8, 676-689.

Catani, M. V., Savini, I., Duranti, G., Caporossi, D., Ceci, R., Sabatini, S. and Avigliano, L. (2004) Nuclear factor kappaB and activating protein 1 are involved in differentiation-related resistance to oxidative stress in skeletal muscle cells. *Free Radic. Biol. Med.*, 37, 1024-1036.

Chakravarthy, M. V., Davis, B. S. and Booth, F. W. (2000) IGF-I restores satellite cell proliferative potential in immobilized old skeletal muscle. J. *Appl. Physiol.*, 89, 1365- 1379.

Charvet, C., Houbron, C., Parlakian, A., Giordani, J., Lahoute, C., Bertrand, A., Sotiropoulos, A., Renou, L., Schmitt, A., Melki, J., Li, Z., Daegelen, D. and Tuil, D. (2006) New role for serum response factor in postnatal skeletal muscle growth and regeneration via the interleukin 4 and insulin-like growth factor 1 pathways. *Mol. Cell. Biol.*, 26, 6664-6674.

Chen, S. E., Gerken, E., Zhang, Y., Zhan, M., Mohan, R. K., Li, A. S., Reid, M. B. and Li, Y. P. (2005) Role of TNF-α signaling in a regeneration of cardiotoxin-injured muscle. *Am. J. Physiol. Cell Physiol.*, 289, C1179-C1187.

Clemmons, D. R. (2009) Role of IGF-I in skeletal muscle mass maintenance. *Trends Endocrinol. Metabol.*, 20, 349-356.

Conboy, I. M. and Rando, T. A. (2002) The regulation of Notch signaling controls satellite cell activation and cell fate determination in postnatal myogenesis. *Dev. Cell.*, 3, 397-409.

Conboy, I. M., Conboy, M. J., Smythe, G. M. and Rando, T. A. (2003) Notch-mediated restoration of regenerative potential to aged muscle. *Science,* 302, 1575-1577.

Conboy, I. M., Conboy, M. J., Wagners, A. J., Girma, E. R., Weissman, I. L. and Rando, T. A. (2005) Rejuvenation of aged progenitor cells by exposure to a young systemic environment. *Nature,* 433, 760-764.

Cornelison, D. D., Olwin, B. B., Rudnicki, M. A. and Wold, B. J. (2000) MyoD(−/−) satellite cells in single-fiber culture are differentiation defective and MRF4 deficient. *Dev. Biol.,* 224, 122-137.

Cosgrove, B., Sacco, A., Gilbert, P. M. and Blau, H. M. (2009) A home away from home: Challenges and opportunities in engineering in vitro muscle satellite cell niches. *Differentiation*, 78, 185-194.

Cross, D. A., Alessi, D. R., Cohen, P., Andjelkovich, M. and Hemmings, B. A. (1995) Inhibition of glycogen synthase kinase-3 by insulin mediated by protein kinase B. *Nature,* 378, 785-789.

Cuenda, A. and Cohen, P. (1999) Stress-activated protein kinase-2/p38 and a rapamycin- sensitive pathway are required for C2C12 myogenesis. *J. Biol. Chem.*, 274, 4341-4346.

Cusella-De Angelis, M. G., Molinari, S., Le Donne, A., Coletta, M., Vivarelli, E., Bouche, M., Molinaro, M., Ferrari, S. and Cossu, G. (1994) Differential response of embryonic and fetal myoblasts to TGF beta: A possible regulatory mechanism of skeletal muscle histogenesis. *Development,* 120, 925-933.

Czifra, G., Tóth, I. B., Marincsák, R., Juhász, I., Kovács, I., Acs, P., Kovács, L., Blumberg, P. M. and Bíró, T. (2006) Insulin-like growh factor-I-coupled mitogenic signaling in primary cultured human skeletal muscle cells and in C2C12 myoblasts. *Cell. Signal.*, 18, 1461- 1472.

Dahlman, J. M., Wang, J., Bakkar, N. and Guttridge, D. C. (2009) The RelA/p65 subunit of NF-kappaB specifically regulates cyclin D1 protein stability: Implications for cell cycle withdrawal and skeletal myogenesis. *J. Cell Biochem.*, 106, 42-51.

Delling, U., Tureckova, J., Lim, H. W., De Windt, L. J., Rotwein, P. and Molkentin, J. D. (2000) A calcineurin-NFATc3-dependent pathway regulates skeletal muscle differentiation and slow myosin heavy-chain expression. *Mol. Cell. Biol.*, 20, 6600-6611.

Devol, D. L., Rotwein, P., Sadow, J. L., Novakofski, J. and Bechtel, P. J. (1990) Activation of insulin-like growth factor gene expression during work-induced skeletal muscle growth. *Am. J. Physiol.*, 259, E89-E95.

Diao, Y., Wang, X. and Wu, Z. (2009) SOCS1, SOCS3, and PIAS1 promote myogenic differentiation by inhibiting the leukemia inhibitory factor-induced JAK1/STAT1/ STAT3 pathway. *Mol. Cell. Biol.*, 29, 5084-5093.

Dijkstra, C. D., Döpp, E. A., Joling, P. and Kraal, G. (1985) The heterogeneity of mononuclear phagocytes in lymphoid organs: Distinct macrophage subpopulations in the rat recognized by monoclonal antibodies ED1, ED2 and ED3. *Immunology*, 54, 589-599.

Dogra, C., Changotra, H., Mohan, S. and Kumar, A. (2006) Tumor necrosis factor-like weak inducer of apoptosis inhibits skeletal myogenesis through sustained activation of nuclear factor-kappaB and degradation of MyoD protein. *J. Biol. Chem.*, 281, 10327- 10336.

Dogra, C., Changotra, H., Wedhas, N., Qin, X., Wergedal, J. E. and Kumar, A. (2007a) TNF-related weak inducer of apoptosis (TWEAK) is a potent skeletal muscle-wasting cytokine. *FASEB J.,* 21, 1857-1869.

Dogra, C., Hall, S. L., Wedhas, N., Linkhart, T. A. and Kumar, A. (2007b) Fibroblast growth factor inducible 14 (Fn14) is required for the expression of myogenic regulatory factors and differentiation of myoblasts into myotubes. Evidence for TWEAK-independent functions of Fn14 during myogenesis. *J. Biol. Chem.*, 282, 15000-15010.

Drummond, M. J., Fry, C. S., Glynn, E. L., Dreyer, H. C., Dhanani, S., Timmerman, K. L., Volpi, E. and Rasmussen, B. B. (2009) Rapamycin administration in humans blocks the contraction-induced increase in skeletal muscle protein synthesis. *J. Physiol.*, 587, 1535-1546.

Erbay, E. and Chen, J. (2001) The mammalian target of rapamycin regulates C2C12 myogenesis via a kinase-independent mechanism. *J. Biol. Chem.*, 276, 36079-36082.

Franchini, K. G., Torsoni, A. S., Soares, P. H. and Saad, M. J. (2000) Early activation of the multicomponent signaling complex associated with focal adhesion kinase induced by pressure overload in the rat heart. *Circ. Res.*, 87, 558-565.

Frenette, J., Cai, B. and Tidball, J. G. (2000) Complement activation promotes muscle inflammation during modified muscle use. *Am. J. Pathol.*, 156, 2103-2110.

Friday, B. B., Horsley, V. and Pavlath, G. K. (2000) Calcineurin activity is required for the initiation of skeletal muscle differentiation. *J. Cell Biol.*, 149, 657-666.

Friday, B. B., Mitchell, P. O., Kegley, K. M. and Pavlath, G. K. (2003) Calcineurin initiates skeletal muscle differentiation by activating MEF2 and MyoD. *Differentiation,* 71, 217-227.

Gauthier-Rouviére, C., Vandromme, M., Tuil, D., Lautredou, N., Morris, M., Soulez, M., Kahn, A., Fernandez, A. and Lamb, N. (1996) Expression and activity of serum response factor is required for expression of the muscle-determining factor MyoD in both dividing and differentiating mouse C2C12 myoblasts. *Mol. Biol. Cell*, 7, 719-729.

Gayraud-Morel, B., Chretien, F., Flamant, P., Gomes, D., Zammit, P. S. and Tajbakhsh, A. (2007) A role for the myogenic determination gene Myf5 in adult regenerative myogenesis. *Dev. Biol.*, 312, 13-28.

Ge, X., McFarlane, C., Vajjala, A., Lokireddy, S., Ng, Z. H., Tan, C. K., Tan, N. S., Wahli, W., Sharma, M. and Kambadur, R. (2011) Smad3 signaling is required for satellite cell function and myogenic differentiation of myoblasts. *Cell Res.*, 21, 1591-1604.

Ge, X., Vajjala, A., McFarlane, C., Wahli, W., Sharma, M. and Kambadur, R. (2012) Lack of Smad3 signaling leads to impaired skeletal muscle regeneration. *Am. J. Physiol. Endocrinol. Metab.*, 303, E90-E102.

Girgenrath, M., Weng, S., Kostek, C. A., Browning, B., Wang, M., Brown, S. A., Winkles, J. A., Michaelson, J. S., Allaire, N., Schneider, P., Scott, M. L., Hsu, Y. M., Yagita, H., Flavell, R. A., Miller, J. B., Burkly, L. C. and Zheng, T. S. (2006) TWEAK, via its receptor Fn14, is a novel regulator of mesenchymal progenitor cells and skeletal muscle regeneration. *EMBO J.*, 25, 5826-5839.

Gordon, S. (2003) Alternative activation of macrophages. *Nat. Rev. Immunol.*, 3, 23-35.

Gordon, S. and Taylor, P. R. (2005) Monocyte and macrophage heterogeneity. *Nat. Rev. Immunol.*, 5, 953-964.

Gosselin, L. E. and McCormick, K. M. (2004) Targeting the immune system to improve ventilatory function in muscular dystrophy. *Med. Sci. Sports Exerc.,* 36, 44-51.

Gregorevic, P., Williams, D. A. and Lynch, G. S. (2002) Effects of leukemia inhibitory factor on rat skeletal muscles are modulated by clenbuterol. *Muscle Nerve*, 25, 194-201.

Guerci, A., Lahoute, C., Hébrard, S., Collard, L., Graindorge, D., Favier, M., Cagnard, N., Batonnet-Pichon, S., Précigout, G., Garcia, L., Tuil, D., Daegelen, D. and Sotiropoulos, A. (2012) Srf-dependent paracrine signals produced by myofibers control satellite cell-mediated skeletal muscle hypertrophy. *Cell Metab*., 15, 25-37.

Guttridge, D. C., Albanese, C., Reuther, J. Y., Pestell, R. G. and Baldwin, A. S. Jr. (1999) NF-kappaB controls cell growth and differentiation through transcriptional regulation of cyclin D1. *Mol. Cell. Biol*., 19, 5785-5799.

Guttridge, D. C., Mayo, M. W., Madrid, L. V., Wang, C. Y. and Baldwin, A. S. Jr. (2000) NF-κB-induced loss of MyoD messenger RNA: Possible role in muscle decay and cachexia. *Science,* 289, 2363-2366.

Guttridge, D. C. (2004) Signaling pathways weigh in on decisions to make or break skeletal muscle. *Curr. Opin. Clin. Nutr. Metab. Care*, 7, 443-450.

Han, J., Jiang, Y., Li, Z., Kravchenko, V. V. and Ulevitch, R. J. (1997) Activation of the transcription factor MEF2C by the MAP kinase p38 in inflammation. *Nature*, 386, 296-299.

Haq, S., Kilter, H., Michael, A., Tao, J., O'Leary, E., Sun, X. M., Walters, B., Bhattacharya, K., Chen, X., Cui, L., Andreucci, M., Rosenzweig, A., Guerrero, J. L., Patten, R., Liao, R., Molkentin, J., Picard, M., Bonventre, J. V. and Force, T. (2003) Deletion of cytosolic phospholipase A2 promotes striated muscle growth. *Nat. Med*., 9, 944-951.

Hawke, T. J. and Garry, D. J. (2001) Myogenic satellite cells: Physiology and molecular biology. *J. Appl. Physiol*., 91, 534-551.

Hayden, M. S. and Ghosh, S. (2008) Shared principles in NF-kappaB signaling. *Cell,* 132, 344-362.

Hinds, M. G., Mauer, T., Zhang, J. G., Nicola, N. A. and Norton, R. S. (1997) Resonance assignments, secondary structure and topology of leukaemia inhibitory factor in solution. *J. Biomed. NMR*., 9, 113-126.

Horsley, V., Friday, B. B., Matteson, S., Kegley, K. M., Gephart, J. and Pavlath, G. K. (2001) Regulation of the growth of multinucleated muscle cells by an NFATC2-dependent pathway. *J. Cell Biol*., 153, 329-338.

Horsley, V., Jansen, K. M., Mills, S. T. and Pavlath, G. K. (2003) IL-4 acts as a myoblast recruitment factor during mammalian muscle growth. *Cell,* 113, 483-494.

Hunt, L. C., Tudor, E. M. and White, J. D. (2010) Leukemia inhibitory factor-dependent increase in myoblast cell number is associated with phosphatidylinositol 3-kinase-mediated inhibition of apoptosis and not mitosis. *Exp. Cell Res.*, 316, 1002-1009.

Jacinto, E., Loewith, R., Schmidt, A., Lin, S., Rüegg, M. A., Hall, A. and Hall, M. N. (2004) Mammalian TOR complex 2 controls the actin cytoskeleton and its rapamycin insensitive. *Nat. Cell Biol.,* 6, 1122-1128.

Jansen, K. M. and Pavlath, G. K. (2008) Molecular control of mammalian myoblast fusion. *Methods Mol. Biol.*, 475, 115-133.

Johnson, S. E. and Allen, R. E. (1995) Activation of skeletal satellite cells and the role of fibroblast growth-factor receptors. *Exp. Cell Res.*, 219, 449-453.

Joulia, D., Bernardi, H., Garandel, V., Rabenoelina, F., Vernus, B. and Cabello, G. (2003) Mechanisms involved in the inhibition of myoblast proliferation and differentiation by myostatin. *Exp. Cell Res.*, 286, 263-275.

Joulia-Ekaza, D. and Cabello, G. (2007) The myostatin gene: Physiology and pharmacological relevance. *Curr. Opin. Pharmacol.*, 7, 310-315.

Kami, K. and Semba, E. (1998) Localization of leukemia inhibitory factor and interleukin-6 messenger ribonucleic acids in regenerating rat skeletal muscle. *Muscle Nerve*, 21, 819-822.

Kami, K., Morikawa, Y., Sekimoto, M. and Senba, E. (2000) Gene expression of receptors for IL-6, LIF, and CNTF in regenerating skeletal muscles. *J. Histochem. Cytochem.*, 48, 1203-1213.

Kielbasa, O. M., Reynolds, J. G., Wu, C. L., Snyder, C. M., Cho, M. Y., Weiler, H., Kandarian, S. and Naya, F. J. (2011) Myospryn is a calcineurin-interacting protein that negatively modulates slow-fiber-type transformation and skeletal muscle regeneration. *FASEB J.*, 25, 2276-2286.

Kitamoto, T. and Hanaoka, K. (2010) Notch3 null mutation in mice causes muscle hyperplasia by repetitive muscle regeneration. *Stem Cells*, 28, 2205-2216.

Kitzmann, M., Bonnieu, A., Duret, C., Vernus, B., Barro, M., Laodj-Chevivesse, D., Verdi, J. M. and Carnac, G. (2006) Inhibition of notch signaling induces myotube hypertrophy by recruiting a subpopulation of reserve cells. *J. Cell. Physiol.*, 208, 538- 548.

Koh, T. J., Bryer, S. C., Pucci, A. M. and Sisson, T. H. (2005) Mice deficient in plasminogen activator inhibitor-1 have improved skeletal muscle regeneration. *Am. J. Physiol. Cell Physiol.*, 289, C217-C223.

Koulmann, N., Sanchez, B., N'Guessan, B., Chapot, R., Serrurier, B., Peinnequin, A., Ventura-Clapier, R. and Bigard, X. (2006) The responsiveness of regenerated soleus muscle to pharmacological calcineurin inhibition. *J. Cell. Physiol.*, 208, 116-122.

Kristiansen, M., Graversen, J. H., Jacobsen, C., Sonne, O., Hoffman, H. J., Law, S. K. and Moestrup, S. K. (2001) Identification of the haemoglobin scavenger receptor. *Nature*, 409, 198-201.

Kuang, S., Charge, S. B., Seale, O., Huh, M. and Rudnicki, M. A. (2006) Distinct roles for Pax7 and Pax3 in adult regenerative myogenesis. *J. Cell Biol.*, 172, 103-113.

Kuang, S., Kuroda, K., Le Grand, F. and Rudnicki, M. A. (2007) Asymmetric self-renewal and commitment of satellite stem cells in muscle. *Cell,* 129, 999-1010.

Kuang, S. and Rudnicki, M. A. (2008) The emerging biology of satellite cells and their therapeutic potential. *Trends Mol. Med.*, 14, 82-91.

Kuang, S., Gillespie, M. A. and Rudnicki, M. A. (2008) Niche regulation of muscle satellite cell self-renewal and differentiation. *Cell Stem Cell*, 2, 22-31.

Kurek, J. B., Bower, J. J., Romanella, M., Koentgen, F., Murphy, M. and Austin, L. (1997) The role of leukemia inhibitory factor in skeletal muscle regeneration. *Muscle Nerve*, 20, 815-822.

Kuwahara, K., Barrientos, T., Pipes, G. C., Li, S. and Olson, E. N. (2005) Muscle-specific signaling mechanism that links actin dynamics to serum response factor. *Mol. Cell. Biol.*, 25, 3173-3181.

Kuwahara, K., Teg Pipes, G. C., McAnally, J., Richardson, J. A., Hill, J. A., Bassel-Duby, R. and Olson, E. N. (2007) Modulation of adverse cardiac remodeling by STARS, a mediator of MEF2 signaling and SRF activity. *J. Clin. Invest.*, 117, 1324-1334.

Lai, K. M., Gonzalez, M., Poueymirou, W. T., Kline, W. O., Na, E., Zlotchenko, E., Stitt, T. N., Economides, A. N., Yancopoulos, G. D. and Glass, D. J. 2004. Conditional activation of akt in adult skeletal muscle induces rapid hypertrophy. *Mol. Cell. Biol.*, 24, 9295-9304.

Lange, S., Xiang, F., Yakovenko, A., Vihola, A., Hackman, P., Rostkova, E., Kristensen, J., Brandmeier, B., Franzen, G., Hedberg, B., Gunnarsson, L. G., Hughes, S. M., Marchand, S., Sejersen, T., Richard, I., Edström, L., Ehler, E., Udd, B. and Gautel, M. (2005) The kinase domain of titin controls muscle gene expression and protein turnover. *Science*, 308, 1599-1603.

Langen, R. C., Schols, A. M., Kelders, M. C., Wouters, E. F. and Janssen-Heininger, Y. M. (2001) Inflammatory cytokines inhibit myogenic differentiation through activation of nuclear factor-κB. *FASEB J.,* 15, 1169-1180.

Langen, R. C., Schols, A. M., Kelders, M. C., Van Der Velden, J. L., Wouters, E. F. and Janssen-Heininger, Y. M. (2002) Tumor necrosis factor-alpha inhibits myogenesis through redox-dependent and -independent pathways. *Am. J. Physiol. Cell Physiol.*, 283, C714-C721.

Langen, R. C., Van der Velden, J. L., Schols, A. M., Kelders, M. C., Wouters, E. F. and Janssen-Heininger, Y. M. (2004) Tumor necrosis factor-α inhibits myogenic differentiation through MyoD protein destabilization. *FASEB J.,* 18, 227-237.

Langen, R. C., Schols, A. M., Kelders, M. C., Van der Velden, J. L., Wouters, E. F. and Janssen-Heininger, Y. M. (2006) Muscle wasting and impaired muscle regeneration in a murine model of chronic pulmonary inflammation. *Am. J. Respir. Cell Mol. Biol.*, 35, 689-696.

Langley, B., Thomas, M., Bishop, A., Sharma, M., Gilmour, S. and Kambadur, R. (2002) Myostatin inhibits myoblast differentiation by down-regulating MyoD expression. *J. Biol. Chem.*, 277, 49831-49840.

Lara-Pezzi, E., Winn, N., Paul, A., McCullagh, K., Slominsky, E., Santini, M. P., Mourkioti, F., Sarathchandra, P., Fukushima, S., Suzuki, K. and Rosenthal, N. (2007) A naturally occurring calcineurin variant inhibits FoxO activity and enhances skeletal muscle regeneration. *J. Cell Biol.,* 179, 1205-1218.

Lee, S. J. (2004) Regulation of muscle mass by myostatin. *Annu. Rev. Cell Dev. Biol.*, 20, 61-86.

Le Grand, F., Jones, A. E., Seale, V., Scimè, A. and Rudnicki, M. A. (2009) Wnt7a activates the planar cell polarity pathway to drive the symmetric expansion of satellite stem cells. *Cell Stem Cell*, 4, 535-547.

Lexell, J. (1993) Ageing and human muscle: Observations from Sweden. *Can. J. Appl. Physiol.,* 18, 2-18.

Li, Y., Foster, W., Deasy, B. M., Chan, Y., Prisk, V., Tang, Y., Cummins, J. and Huard, J. (2004) Transforming growth factor-beta1 induces the differentiation of myogenic cells into fibrotic cells in injured skeletal muscle: A key event in muscle fibrogenesis. *Am. J. Pathol.*, 164, 1007-1019.

Linehan, S. A. (2005) The mannose receptor is expressed by subsets of APC in non-lymphoid organs. *BMC Immunol.*, 6, 4.

Lolmede, K., Campana, L., Vezzoli, M., Bosurgi, L., Tonlorenzi, R., Clementi, E., Bianchi, M. E., Cossu, G., Manfredi, A. A., Brunelli, S. and Rovere-Querini, P. (2009) Inflammatory and alternatively activated human macrophages attract vessel-associated stem cells, relying on separate HMGB1- and MMP-9-dependent pathways. *J. Leukoc. Biol.*, 85, 779-787.

Mackey, A. L., Kjaer, M., Dandanell, S., Mikkelsen, K. H., Holm, L., Døssing, S., Kadi, F., Koskinen, S. O., Jensen, C. H., Schrøder, H. D. and Langberg, H. (2007) The influence of anti-inflammatory medication on exercise-induced myogenic precursor cell responses in humans. *J. Appl. Physiol.*, 103, 425-431.

Manning, B. D., Tee, A. R., Logsdon, M. N., Blenis, J. and Cantley, L. C. (2002) Identification of the tuberous sclerosis complex-2 tumor suppressor gene product tuberin as a target of the phosphoinositide 3-kinase/akt pathway. *Mol. Cell*, 10, 151-162.

Mantovani, A., Sica, A., Sozzani, S., Allavena, P., Vecchi, A. and Locati, M. (2004) The chemokine system in diverse forms of macrophage activation and polarization. *Trends Immunol.*, 25, 677-686.

Mantovani, A., Sica, A. and Locati, M. (2007) New vistas on macrophage differentiation and activation. *Eur. J. Immunol.*, 37, 14-16.

Mauro, A. (1961) Satellite cell of skeletal muscle fibers. J. *Biophys. Biochem. Cytol.*, 9, 493-495.

McCarthy, J. J., Mula, J., Miyazaki, M., Erfani, R., Garrison, K., Farooqui, A. B., Srikuea, R., Lawson, B. A., Grimes, B., Keller, C., Van Zant, G., Campbell, K. S., Esser, K. A., Dupont-Versteegden, E. E. and Peterson, C. A. (2011) Effective fiber hypertrophy in satellite cell-depleted skeletal muscle. *Development*, 138, 3657-3666.

McCroskery, S., Thomas, M., Maxwell, L., Sharma, M. and Kambadur, R. (2003) Myostatin negatively regulates satellite cell activation and self-renewal. *J. Cell Biol.*, 162, 1135-1147.

McLennan, I. S. (1993) Localisation of transforming growth factor beta 1 in developing muscles: Implications for connective tissue and fiber type pattern formation. *Dev. Dyn.*, 197, 281-290.

Meadows, S. M., Warkman, A. S., Salanga, M. C., Small, E. M. and Krieg, P. A. (2008) The myocardin-related transcription factor, MASTR, cooperates with MyoD to activate skeletal muscle gene expression. *Proc. Natl. Acad. Sci. U. S. A.*, 105, 1545-1550.

Megeney, L. A., Kablar, B., Garrett, K., Anderson, J. E. and Rudnicki, M. A. (1996) MyoD is required for myogenic stem cell function in adult skeletal muscle. *Genes Dev.*, 10, 1173-1183.

Meighan-Mantha, R. L., Hsu, D. K., Guo, Y., Brown, S. A., Feng, S. L., Peifley, K. A., Alberts, G. F., Copeland, N. G., Gilbert, D. J., Jenkins, N. A., Richards, C. M. and Winkles, J. A. (1999) The mitogen-inducible Fn14 gene encodes a type I transmembrane protein that modulates fibroblast adhesion and migration. *J. Biol. Chem.*, 274, 33166-33176.

Metcalf, D. (2003) The unsolved enigmas of leukemia inhibitory factor. *Stem Cells*, 21, 5-14.

Michel, R. N., Chin, E. R., Chakkalakal, J. V., Eibl, J. K. and Jasmin, B. J. (2007) $Ca^{2+}$/ calmodulin-based signalling in the regulation of the muscle fibre phenotype and its therapeutic potential via modulation of utrophin A and myostatin expression. *Appl. Physiol. Nutr. Metab.*, 32, 921-929.

Milasincic, D. J., Calera, M. R., Farmer, S. R. and Pilch, P. F. (1996) Stimulation of C2C12 myoblast growth by basic fibroblast growth factor and insulin-like growth factor 1 can occur via mitogen-activated protein kinase-dependent and -independent pathways. *Mol. Cell. Biol.*, 16, 5964-5973.

Miller, K. J., Thaloor, D., Matteson, S. and Pavlath, G. K. (2000) Hepatocyte growth factor affects satellite cell activation and differentiation in regenerating skeletal muscle. *Am. J. Physiol. Cell Physiol.*, 278, C174-C181.

Mittal, A., Bhatnagar, S., Kumar, A., Paul, P. K., Kuang, S. and Kumar, A. (2010) Genetic ablation of TWEAK augments regeneration and post-injury growth of skeletal muscle in mice. *Am. J. Pathol.*, 177, 1732-1742.

Moestrup, S. K. and Moller, H. J. (2004) CD163: A regulated hemoglobin scavenger receptor with a role in the anti-inflammatory response. *Ann. Med.,* 36, 347-354.

Mokalled, M. H., Johnson, A. N., Creemers, E. E. and Olson, E. N. (2012) MASTR directs MyoD-dependent satellite cell differentiation during skeletal muscle regeneration. *Genes Dev.*, 26, 190-202.

Morrison, S. and Kimble, J. (2006) Asymmetric and symmetric stem-cell divisions in development and cancer. *Nature*, 441, 1068-1074.

Mosser, D. M. and Edwards, J. P. (2008) Exploring the full spectrum of macrophage activation. *Nat. Rev. Immunol.*, 8, 958-969.

Mourkioti, F., Kratsios, P., Luedde, T., Song, Y. H., Delafontaine, P., Adami, R., Parente, V., Bottinelli, R., Pasparakis, M. and Rosenthal, N. (2006)

Targeted ablation of IKK2 improves skeletal muscle strength, maintains mass, and promotes regeneration. *J. Clin. Invest.*, 116, 2945-2954.

Mozzetta, C., Minetti, G. and Puri, P. L. (2009) Regenerative pharmacology in the treatment of genetic diseases: The paradigm of muscular dystrophy. *Int. J. Biochem. Cell Biol.*, 41, 701-710.

Musaró, A., McCullagh, K., Paul, A., Houghton, L., Dobrowolny, G., Molinaro, M., Barton, E. R., Sweeney, H. L. and Rosenthal, N. (2001) Localized Igf-1 transgene expression sustains hypertrophy and regeneration in senescent skeletal muscle. *Nat. Genet.*, 27, 195-200.

Muscat, G. E., Gustafson, T. A. and Kedes, L. (1988) A common factor regulates skeletal and cardiac alpha-actin gene transcription in muscle. *Mol. Cell. Biol.*, 8, 4120-4133.

Negoro, S., Oh, H., Tone, E., Kunisada, K., Fujio, Y., Walsh, K., Kishimoto, T. and Yamauchi-Takihara, K. (2001) Glycoprotein 130 regulates cardiac myocyte survival in doxorubicin-induced apoptosis through phosphatidylinositol 3-kinase/akt phosphorylation and Bcl-xL/caspase-3 interaction. *Circulation*, 103, 555-561.

Ni, Y. G., Wang, N., Cao, D. J., Sachan, N., Morris, D. J., Gerard, R. D., Kuro-O, M., Rothermel, B. A. and Hill, J. A. (2007) FoxO transcription factors activate Akt and attenuate insulin signaling in heart by inhibiting protein phosphatases. *Proc. Natl. Acad. Sci. U. S. A.*, 104, 20517-20522.

Ochala, J., Gustafson, A. M., Diez, M. L., Renaud, G., Li, M., Aare, S., Qaisar, R., Bauduseela, V. C., Hedström, Y., Tang, X., Dworkin, B., Ford, G. C., Nair, K. S., Perera, S., Gautel, M. and Larsson, L. (2011) Preferential skeletal muscle myosin loss in response to mechanical silencing in a novel rat intensive care unit model: Underlying mechanisms. *J. Physiol.*, 589, 2007-2026.

Ochoa, O., Sun, D., Reyes-Reyna, S. M., Waite, L. L., Michalek, J. E., McManus, L. M. and Shireman, P. K. (2007) Delayed angiogenesis and VEGF production in CCR2-/- mice during impaired skeletal muscle regeneration. *Am. J. Physiol. Regul. Integr. Comp. Physiol.*, 293, R651-R661.

Olguin, H. C., Yang, Z., Tapscott, S. J. and Olwin, B. B. (2007) Reciprocal inhibition between Pax7 and muscle regulatory factors modulates myogenic cell fate determination. *J. Cell Biol.*, 177, 769-779.

Ono, Y., Gnocchi, V. F., Zammit, P. S. and Nagatomi, R. (2009) Presenilin-1 acts via Id1 to regulate the function of muscle satellite cells in a gamma-secretase-independent manner. *J. Cell. Sci.,* 122, 4427-4438.

Ottnad, E., Parthasarathy, S., Sambrano, G. R., Ramprasad, M. P., Quehenberger, O., Kondratenko, N., Green, S. and Steinberg, D. (1995) A macrophage receptor for oxidized low density lipoprotein distinct from the receptor for acetyl low-density lipoprotein: Partial purification and role in recognition of oxidatively damaged cells. *Proc. Natl. Acad. Sci. U. S. A.,* 92, 1391-1395.

Palumbo, R., Galvez, B. G., Pusterla, T., De Marchis, F., Cossu, G., Marcu, K. B. and Bianchi, M. E. (2007) Cells migrating to sites of tissue damage in response to the danger signal HMGB1 require NF-kappaB activation. *J. Cell Biol.*, 179, 33-40.

Park, I. H. and Chen, J. (2005) Mammalian target of rapamycin (mTOR) signaling is required for a late-stage fusion process during skeletal myotube maturation. *J. Biol. Chem.*, 280, 32009-32017.

Pavlath, G. K., Dominov, J. A., Kegley, K. M. and Miller, J. B. (2003) Regeneration of transgenic skeletal muscles with altered timing of expression of the basic helix-loop- helix muscle regulatory factor MRF4. *Am. J. Pahol.*, 162, 1685-1691.

Pelosi, L., Giacinti, C., Nardis, C., Borsellino, G., Rizzuto, E., Nicoletti, C., Wannenes, F., Battistini, L., Rosenthal, N., Molinaro, M. and Musarò, A. (2007) Local expression of IGF-I accelerates muscle regeneration by rapidly modulating inflammatory cytokines and chemokines. *FASEB J.,* 21, 1393-1402.

Peterson, J. M. and Guttridge, D. C. (2008) Skeletal muscle diseases, inflammation, and NF-kappaB signaling: Insights and opportunities for therapeutic intervention. *Int. Rev. Immunol.*, 27, 375-387.

Philippidis, P., Mason, J. C., Evans, B. J., Nadra, I., Taylor, K. M., Haskard, D. O. and Landis, R. C. (2004) Hemoglobin scavenger receptor CD163 mediates interleukin-10 release and heme oxygenase-1 synthesis: Anti-inflammatory monocyte-macrophage responses in vitro, in resolving skin blisters in vivo, and after cardiopulmonary bypass surgery. *Circ. Res.*, 94, 119-126.

Philippou, A., Maridaki, M., Halapas, A. and Koutsilieris, M. (2007) The role of the insulin-like growth factor 1 (IGF-1) in skeletal muscle physiology. *In Vivo,* 21, 45-54.

Pipes, G. C., Creemers, E. E. and Olson, E. N. (2006) The myocardin family of transcriptional coactivators: Versatile regulators of cell growth, migration, and myogenesis. *Genes Dev.,* 20, 1545-1556.

Pisconti, A., Brunelli, S., Di Padova, M., De Palma, C., Deponti, D., Baesso, S., Sartorelli, V., Cossu, G. and Clementi, E. (2006) Follistatin induction by nitric oxide through cyclic GMP: A tightly regulated signaling pathway that controls myoblast fusion. *J. Cell Biol.*, 172, 233-244.

Puchner, E. M., Alexandrovich, A., Kho, A. L., Hensen, U., Schäfer, L. V., Brandmeier, B., Gräter, F., Grubmüller, H., Gaub, H. E. and Gautel, M. (2008) Mechanoenzymatics of titin kinase. *Proc. Natl. Acad. Sci. U. S. A.*, 105, 13385-13390.

Raingeaud, J., Gupta, S., Rogers, J. S., Dickens, M., Han, J., Ulevitch, R. J. and Davis, R. J. (1995) Pro-inflammatory cytokines and environmental stress cause p38 mitogen- activated protein kinase activation by dual phosphorylation on tyrosine and threonine. *J. Biol. Chem.*, 270, 7420-7426.

Reymond, A., Meroni, G., Fantozzi, A., Merla, G., Cairo, S., Luzi, L., Riganelli, D., Zanaria, E., Messali, S., Cainarca, S., Guffanti, A., Minucci, S., Pelicci, P. G. and Ballabio, A. (2001) The tripartite motif family identifies cell compartments. *EMBO J.*, 20, 2140- 2151.

Rommel, C., Bodine, S. C., Clarke, B. A., Rossman, R., Nunez, L., Stitt, T. N., Yancopoulos, G. D. and Glass, D. J. (2001) Mediation of IGF-I-induced skeletal myotube hypertrophy by $PI_3K$/Akt/mTOR and $PI_3K$/Akt/GSK3 pathways. *Nat. Cell Biol.*, 3, 1009-1013.

Roubenoff, R. and Hughes, V. A. (2000) Sarcopenia: Current concepts. *J. Gerontol. A Biol. Sci. Med. Sci.*, 55, M716-M724.

Sabourin, L. A., Girgis-Gabardo, A., Seale, P., Asakura, A. and Rudnicki, M. A. (1999) Reduced differentiation potential of primary MyoD−/− myogenic cells derived from adult skeletal muscle. *J. Cell Biol.*, 144, 631-643.

Sakamoto, K., Hirshman, M. F., Aschenbach, W. G. and Goodyear, L. J. (2002) Contraction regulation of Akt in rat skeletal muscle. *J. Biol. Chem.*, 277, 11910-11917.

Sakuma, K., Watanabe, K., Totsuka, T., Uramoto, I., Sano, M. and Sakamoto, K. (1998) Differential adaptations of insulin-like growth factor-I, basic fibroblast growth factor, and leukemia inhibitory factor in the plantaris muscle of rats by mechanical overloading: An immunohistochemical study. *Acta. Neuropathol.*, 95, 123-130.

Sakuma, K., Watanabe, K., Sano, M., Uramoto, I. and Totsuka, T. (2000) Differential adaptation of growth and differentiation factor 8/myostatin, fibroblast growth factor 6 and leukemia inhibitory factor in overloaded, regenerating, and denervated rat muscles. *Biochim. Biophys. Acta Mol. Cell Res.*, 1497, 77-88.

Sakuma, K., Nishikawa, J., Nakao, R., Watanabe, K., Totsuka, T., Nakano, H., Sano, M. and Yasuhara, M. (2003a) Calcineurin is a potent regulator for skeletal muscle regeneration by association with NFATc1 and GATA-2. *Acta Neuropathol.*, 105, 271-280.

Sakuma, K., Nishikawa, J., Nakao, R., Nakano, H., Sano, M. and Yasuhara, M. (2003b) Serum response factor plays an important role in the mechanically overloaded plantaris muscle of rats. *Histochem. Cell Biol.*, 119, 149-160.

Sakuma, K., Nakao, R., Aoi, W., Inashima, S., Fujikawa, T., Hirata, M., Sano, M. and Yasuhara, M. (2005) Cyclosporin A treatment upregulates Id1 and Smad3 expression and delays skeletal muscle regeneration. *Acta Neuropathol.*, 110, 269-280.

Sakuma, K., Akiho, M., Nakashima, H., Akima, H. and Yasuhara, M. (2008) Age-related reductions in expression of serum response factor and myocardin-related transcription factor A in mouse skeletal muscles. *Biochim. Biophys. Acta Mol. Basis Dis.*, 1782, 453-461.

Sakuma, K. and Yamaguchi, A. (2010) Molecular mechanisms in aging and current strategies to counteract sarcopenia. *Curr. Aging Sci.*, 3, 90-101.

Sakuma, K. and Yamaguchi, A. (2011a) Sarcopenia: Molecular mechanisms and current therapeutic strategy. In: Perloft, J. W. and Wong, A. H., Eds. *Cell Aging*. Nova Science Publishers, NY, pp 93-152.

Sakuma, K. and Yamaguchi, A. (2011b) Serum response factor (SRF)-dependent pathway: Potential mediators of growth, regeneration, and hypertrophy of skeletal muscle. In: Pandalai, S. G., Eds. *Recent Res. Devel. Life Sci., 5*. Research Signpost, Kerala, India, pp. 13-37.

Sakuma, K. and Yamaguchi, A. (2012) Sarcopenia and age-related endocrine function. *Int. J. Endocrinol.*, 2012, Article ID 127362, 10 pages.

Sarbassov, D. D., Ali, S. M. and Sabatini, D. M. (2005) Growing roles for the mTOR pathway. *Curr. Opin. Cell Biol.*, 17, 596-603.

Sardiello, M., Cairo, S., Fontanella, B., Ballabio, A. and Meroni, G. (2008) Genomic analysis of the TRIM family reveals two groups of genes with distinct evolutionary properties. *BMC Evol. Biol.*, 8, 225.

Sarparanta, J. (2008) Biology of myospryn: What's known? *J. Muscle Res. Cell Motil.*, 29, 177-180.

Sartori, R., Milan, G., Patron, M., Mammucari, C., Blaauw, B., Abraham, R. and Sandri, M. (2009) Smad2 and 3 transcription factors control muscle mass in adulthood. *Am. J. Physiol. Cell Physiol.*, 296, C1248-C1257.

Schmelzer, C. H., Burton, L. E. and Tamony, C. M. (1990) Purification and partial characterization of recombinant human differentiation-stimulating factor. *Protein Expr. Purif.*, 1, 54-62.

Scott, D., Blizzard, L., Fell, J. and Jones, G. (2010) The epidemiology of sarcopenia in community living older adults: What role does lifestyle play? *J. Cachex. Sarcopenia Muscle*, 2, 125-134.

Semsarian, C., Wu, M. -J., Ju, Y. -K., Marciniec, T., Yeoh, T., Allen, D. G., Harvey, R. P. and Graham, R. M. (1999) Skeletal muscle hypertrophy is mediated by a $Ca^{2+}$-dependent calcineurin signaling pathway. *Nature*, 400, 576-581.

Serrano, A. L., Baeza-Raja, B., Perdiguero, E., Jardi, M. and Munoz-Canoves, P. (2008) Interleukin-6 is an essential regulator of satellite cell-mediated skeletal muscle hypertrophy. *Cell Metab.*, 7, 33-44.

Serrano, A. L. and Munoz-Canoves, P. (2010) Regulation and dysregulation of fibrosis in skeletal muscle. *Exp. Cell Res.,* 316, 3050-3058.

Skapek, S. X., Rhee, J., Spicer, D. B. and Lassar, A. B. (1995) Inhibition of myogenic differentiation in proliferating myoblasts by cyclin D1-dependent kinase. *Science*, 267, 1022-1024.

Soulez, M., Rouviere, C. G., Chafey, P., Hentzen, D., Vandromme, M., Lautredou, N., Lamb, N., Kahn, A. and Tuil, D. (1996) Growth and differentiation of C2 myogenic cells are dependent on serum response factor. *Mol. Cell. Biol.*, 16, 6065-6074.

Spencer, J. A., Eliazer, S., Ilaria, R. L. Jr., Richardson, J. A. and Olson, E. N. (2000) Regulation of microtubule dynamics and myogenic dfferentiation by MuRF, a striated muscle RING-finger protein. *J. Cell Biol.*, 150, 771-784.

Stoick-Cooper, C. L., Moon, R. T. and Weidinger, G. (2007) Advances in signaling in vertebrate regeneration as a preclude to regenerative medicine. *Genes Dev.*, 21, 1292- 1315.

St Pierre, B. A. and Tidball, J. G. (1994) Differential response of macrophage subpopulations to soleus muscle reloading after rat hindlimb suspension. *J. Appl. Physiol.*, 77, 290-297.

Strle, K., Broussard, S. R., McCusker, R. H., Shen, W. H., LeCleir, J. M., Johnson, R. W., Freund, G. G., Dantzer, R. and Kelley, K. W. (2006) C-jun N-terminal kinase mediates tumor necrosis factor-alpha suppression of differentiation in myoblasts. *Endocrinology,* 147, 4363-4373.

Strle, K., McCusker, R. H., Tran, L., King, A., Johnson, R. W., Freund, G. G., Dantzer, R. and Kelley, K. W. (2007) Novel activity of an anti-inflammatory cytokine: IL-10 prevents TNFalpha-induced resistance to IGF-I in myoblasts. *J. Neuroimmunol.*, 188, 48-55.

Stupka, N., Schertzer, J. D., Bassel-Duby, R., Olson, E. N. and Lynch, G. S. (2007) Calcineurin-Aα activation enhances the structure and function of regenerating muscles after myotoxic injury. *Am. J. Physiol. Regul. Integr. Comp. Physiol.*, 293, R686-R694.

Suelves, M., Lluís, F., Ruiz, V., Nebreda, A. R., Chapman, R., Hulderman, T., Van Rooijen, N. and Simeonova, P. P. (2004) Phosphorylation of MRF4 transactivation domain by p38 mediates repression of specific myogenic genes. *EMBO J.,* 23, 365-375.

Sun, L., Ma, K., Wang, H., Xiao, F., Gao, Y., Zhang, W., Wang, K., Gao, X., Ip, N. and Wu, Z. (2007) JAK1-STAT1-STAT3, a key pathway promoting proliferation and preventing premature differentiation of myoblasts. *J. Cell Biol.*, 179, 129-138.

Sun, Y., Ge, Y., Drnevich, J., Zhao, Y., Band, M. and Chen, J. (2010) Mammalian target of rapamycin regulates miRNA-1 and follistatin in skeletal myogenesis. *J. Cell Biol.*, 189, 1157-1169.

Tajbakhsh, S. (2009) Skeletal muscle stem cells in developmental versus regenerative myogenesis. *J. Intern. Med.*, 266, 372-389.

Tatsumi, R., Anderson, J. E., Nevoret, C. J., Halevy, O., and Allen, R. E. (1998) HGF/SF is present in normal adult skeletal muscle and is capable of activating satellite cells. *Dev. Biol.*, 194, 114-128.

Tatsumi, R., Liu, X., Pulido, A., Morales, M., Sakata, T., Dial, S., Hattori, A., Ikeuchi, Y. and Allen, R. E. (2006) Satellite cell activation in stretched skeletal muscle and the role of nitric oxide and hepatocyte growth factor. *Am. J. Physiol. Cell Physiol.*, 290, C1487-C1494.

Tidball, J. G., Lavergne, E., Lau, K. S., Spencer, M. J., Stull, J. T. and Wehling, M. (1998) Mechanical loading regulates NOS expression and activity in developing and adult skeletal muscle. *Am. J. Physiol. Cell Physiol.*, 275, C260-C266.

Tidball, J. G. and Wehling-Henricks, M. (2007) Macrophages promote muscle membrane repair and muscle fibre growth and regeneration during modified muscle loading in mice in vivo. *J. Physiol.*, 578, 327-336.

Tiffin, N., Adi, S., Stokoe, D., Wu, N. Y. and Rosenthal, S. M. (2004) Akt phosphorylation is not sufficient for insulin-like growth factor-stimulated myogenin expression but must be accompanied by down-regulation of mitogen-activated protein kinase/extracellular signal-regulated kinase phosphorylation. *Endocrinology*, 145, 4991-4996.

Treisman, R. (1987) Identification and purification of a polypeptide that binds to the c-fos serum response element. *EMBO J.,* 6, 2711-2717.

Tsujinaka, T., Fujita, J., Ebisui, C., Yano, M., Kominami, E., Suzuki, K., Tanaka, K., Katsume, A., Ohsugi, Y., Shiozaki, H. and Monden, M. (1996) Interleukin 6 receptor antibody inhibits muscle atrophy and modulates proteolytic systems in interleukin 6 transgenic mice. *J. Clin. Invest.*, 97, 244-249.

Tureckova, J., Wilson, E. M., Cappalonga, J. L. and Rotwein, P. (2001) Insulin-like growth factor-mediated muscle differentiation: Collaboration between phosphatidylinositol 3-kinase-Akt-signaling pathways and myogenin. *J. Biol. Chem.*, 276, 39264-39270.

Van Amerongen, R. and Nusse, R. (2009) Towards an integrated view of Wnt signaling in development. *Development*, 136, 3205-3214.

Vasyutina, E., Lenhard, D. C., Wende, H., Erdmann, B., Epstein, J. A. and Birchmeier, C. (2007) RBP-J (Rbpsuh) is essential to maintain muscle progenitor cells and to generate satellite cells. *Proc. Natl. Acad. Sci. U. S. A.*, 104, 4443-4448.

Villalta, S. A., Nguyen, H. X., Deng, B., Gotoh, T. and Tidball, J. G. (2009) Shifts in macrophage phenotypes and macrophage competition for arginine metabolism affect the severity of muscle pathology in muscular dystrophy. *Hum. Mol. Genet.*, 18, 482-496.

Wang, H., Hertlein, E., Bakkar, N., Sun, H., Acharyya, S., Wang, J., Carathers, M., Davuluri, R. and Guttridge, D. C. (2007) NF-kappaB regulation of YY1 inhibits skeletal myogenesis through transcriptional silencing of myofibrillar genes. *Mol. Cell. Biol.*, 27, 4374-4387.

Wang, X., Wu, H., Zhang, Z., Liu, S., Yang, J., Chen, X. and Fan, M. (2008) Effects of interleukin-6, leukemia inhibitory factor, and ciliary neurotrophic factor on the proliferation and differentiation of adult human myoblasts. *Cell. Mol. Neurobiol.*, 28, 113-124.

Warren, G. L., Hulderman, T., Jensen, N., McKinstry, M., Mishra, M., Luster, M. I. and Simoneva, P. P. (2002) Physiological role of tumor necrosis factor α in a traumatic muscle injury. *FASEB J.,* 16, 1630-1632.

White, J. D., Davies, M. and Grounds, M. D. (2001) Leukemia inhibitory factor increases myoblast replication and survival and affects extracellular matrix production: Combined in vivo and in vitro studies in postnatal skeletal muscle. *Cell Tissue Res.*, 306, 129-141.

Winkles, J. A., Tran, N. L., Brown, S. A., Stains, N., Cunliffe, H. E. and Berens, M. E. (2007) Role of TWEAK and Fn14 in tumor biology. *Front. Biosci.*, 12, 2761-2771.

Winkles, J. A. (2008) TWEAK-Fn14 cytokine-receptor axis: Discovery, biology and therapeutic targeting. *Nat. Rev. Drug Discov.*, 7, 411-425.

Wolfman, N. M., McPherron, A. C., Pappano, W. N., Davies, M. V., Song, K., Tomkinson, K. N., Wright, J. F., Zhao, L., Sebald, S. M., Greenspan, D. S. and Lee, S. J. (2003) Activation of latent myostatin by the BMP-1/tolloid family of metalloproteinases. *Proc. Natl. Acad. Sci. U. S. A.,* 100, 15842-15846.

Wu, Z., Woodring, P. J., Bhakta, K. S., Tamura, K., Wen, F., Feramisco, J. R., Karim, M., Wang, J. Y. and Puri, P. L. (2000) p38 and extracellular signal-regulated kinases regulate the myogenic program at multiple steps. *Mol. Cell. Biol.*, 20, 3951-3964.

Wullschleger, S., Loewith, R. and Hall, M. N. (2006) TOR signaling in growth and metabolism. *Cell*, 124, 471-484.

Yamada, M., Tatsumi, R., Yamanouchi, K., Hosoyama, T., Shiratsuchi, S., Sato, A., Mizunoya, W., Ikeuchi,Y., Furuse, M. and Allen, R. E. (2010) High concentrations of HGF inhibit skeletal muscle satellite cell proliferation in vitro by inducing expression of myostatin: A possible mechanism for reestablishing satellite cell quiescence in vivo. *Am. J. Physiol. Cell Physiol.*, 298, C465-C476.

Yang, W., Zhang, Y., Li, Y., Wu, Z. and Zhu, D. (2007) Myostatin induces cyclin D1 degradation to cause cell cycle arrest through a phosphatidylinositol 3-kinase/AKT/GSK-3β pathway and is antagonized by insulin-like growth factor 1. *J. Biol. Chem.*, 282, 3799- 3808.

Zádor, E., Mendler, L., Takács, V., De Bleecker, J. and Wuytack, F. (2001) Regenerating soleus and extensor digitorum longus muscles of the rat show elevated levels of TNF-α and its receptors, TNFR-60 and TNFR-80. *Muscle Nerve*, 21, 1058-1067.

Zammit, P. S., Relaix, F., Nagata, Y., Ruiz, A. P., Collins, C. A., Partridge, T. A. and Beauchamp, J. R. (2006) Pax7 and myogenic progression in skeletal muscle satellite cells. *J. Cell. Sci.,* 119, 1824-1832.

Zetser, A., Gredinger, E. and Bengal, E. (1999) p38 mitogen-activated protein kinase pathway promotes skeletal muscle differentiation. Participation of the Mef2c transcription factor. *J. Biol. Chem.*, 274, 5193-5200.

Zimmers, T. A., Davies, M. V., Koniaris, L. G., Haynes, P., Esquela, A. F., Tomkinson, K. N., McPherron, A. C., Wolfman, N. M. and Lee, S. J. (2002) Induction of cachexia in mice by systemically administerd myostatin. *Science,* 296, 1486-1488.

Zoncu, R., Efeyan, A. and Sabatini, D. M. (2011) mTOR: From growth signal integration to cancer, diabetes and ageing. *Nat. Rev. Mol. Cell Biol.,* 12, 21-35.

In: Muscle Cells
Editor: Benigno Pezzo

ISBN: 978-1-62417-233-5


*Chapter 2*

# CALVARIAL AND PERIODONTAL TISSUE INDUCTION BY AUTOGENOUS STRIATED MUSCLE STEM CELLS

***Ugo Ripamonti*[*], *Ansuyah Magan, Roland M. Klar and June Teare***

Bone Research Laboratory, Faculty of Health Sciences,
School of Physiology, University of the Witwatersrand,
Johannesburg, South Africa

## ABSTRACT

The central question in developmental biology, tissue engineering and regenerative medicine at large, is the molecular basis of pattern formation, tissue induction and morphogenesis. The three requirements for the induction of tissue morphogenesis are a suitable biomimetic extracellular matrix substratum, soluble inductive molecular signals, and responding stem cells capable of ligating soluble molecular signals. Tissue induction and morphogenesis by combinatorial molecular protocols is epitomized by the sequential cascade of "*Bone: Formation by autoinduction*". Any of the three variables in the equation can be modified and modulated to initiate the induction of bone formation in

[*] Corresponding author: Bone Research Laboratory, Medical School, & York Road, 2193 Parktown, South Africa; Tel: +27 11 717 2144; Fax: +27 11 717 2300; E-mail: ugo.ripamonti@wits.ac.za

skeletal defects of the craniofacial and appendicular skeletons. A number of isoforms of soluble osteogenic molecular signals may be recombined or reconstituted with different extracellular matrix substrata to biomimetize the structure/activity profile of the extracellular matrix as well as of the osteogenic soluble molecular signals. Stem cells with selected ligands' receptors are capable of differentiating and inducing selected tissue phenotypes and morphogenesis. Progenitor stem cells are either locally stimulated by available soluble molecular signals or can be additionally isolated and intra-operatively added to the surgical site providing an adjunctive tool to therapeutic bone tissue engineering. Striated muscle represents an abundant source of easily accessible tissue that contains several perivascular and intramuscular cell niches available for tissue engineering applications. Myoblastic stem cells including myoendothelial stem cells harvested from striated muscle represent a therapeutic advancement in regenerative medicine and tissue engineering for craniofacial and periodontal applications. Muscular tissue also contains mesenchymal stem cells now known to be pericytes attached to perivascular niches. Morcellated fragments of autogenous *rectus abdominis* muscle containing large quantities of pericytes and myoendothelial cells delivered by Matrigel® matrix and insoluble collagenous bone matrix recombined with recombinant human transforming growth factor-$\beta_3$ (hTGF-$\beta_3$) enhance calvarial regeneration in the non-human primate *Papio ursinus*. Morcellated fragments of autogenous *rectus abdominis* muscle combined with soluble osteogenic molecular signals induce greater amounts of alveolar bone and cementum regeneration along the exposed root surfaces in periodontal defects of *Papio ursinus*. Importantly, morcellated fragments of striated muscle are relatively surgically accessible not only from the *rectus abdominis* but from the orofacial muscular tissues. Harvested fragments require minimal surgical preparation and none *in vitro*, yet retain significant regenerative potential directed by the surrounding extracellular matrices, i.e. osteogenic in craniofacial osseous sites and cementogenic when in contact with dentine extracellular matrices.

**Keywords:** primates, *rectus abdominis* myoblastic stem cells, osteogenic proteins of TGF-β supergene family, calvarial defects, periodontal furcation defects, tissue induction and regeneration, cementogenesis

## Combinatorial Molecular Protocols and the Induction of Bone Formation

Regenerative medicine is the grand multidisciplinary challenge of molecular, cellular and evolutionary biology requiring the integration of tissue biology, tissue engineering, developmental biology and experimental surgery to explore how to trigger *de novo* and *ex novo* tissue induction and morphogenesis to generate new tissues and organs in man (Reddi 1994; Reddi 2000).

The rapidly emerging question in tissue engineering and regenerative medicine at large is whether biomaterial matrices designed for tissue induction and morphogenesis could be additionally assembled with muscular/myoblastic stem cells to further enhance tissue induction and morphogenesis in clinical contexts. Several studies have shown that, in general, the addition of stem cells and/or progenitors, as well as fully differentiated osteoblastic cell lines result in superior tissue induction at the site of surgical implantation in pre-clinical contexts (Usas and Huard 2007; Lee *et al.*, 2000). The question still remains, however, whether such rather sophisticated cell isolation and cloning techniques could be routinely deployed in human patients avoiding the possible alloantigenic load but particularly the costs involved in such highly sophisticated *in vitro* and *ex vivo* laboratory and surgical procedures.

This chapter reports a series of regenerative procedures in the non-human primate Chacma baboon *Papio ursinus* as a prerequisite for potential novel therapeutic strategies in clinical contexts. These would deploy myoblastic/myoendothelial and pericytic/perivascular stem cells all contained in morcellated fragments of autogenous *rectus abdominis* muscle transplanted together with different vehicolating carriers in calvarial and periodontal furcation defects in *Papio ursinus*. The equation of the tissue engineering paradigm is thus manipulated by the addition of large amounts of autologous myoblastic responding stem cells which are implanted in surgically created defects. The induction of bone formation has been shown to be positively affected. This is of interest, since the preparation of autogenous morcellated fragments of *rectus abdominis* muscle does not require laboratory *in vitro* manipulations such as isolation and cloning techniques. Biopsied muscular fragments only require harvesting, and when morcellated, retain high regenerative capacities when implanted in calvarial and periodontal alveolar bone defects (Ripamonti *et al.*, 2008; Ripamonti *et al.*, 2009c; Ripamonti *et al.*, 2009a).

Striated muscle has shown the presence of several stem cell niches (Usas and Huard 2007); amongst muscle derived stem cells (MDSCs) other myoblastic and perivascular pericytic stem cells, the latter now known to be the ubiquitous mesenchymal stem cells (MSCs), the archetypal multipotent progenitor cells developed in culture of developed organs (Crisan *et al.*, 2008). Further work has shown that skeletal muscle cells contain osteoprogenitor cells (Usas and Huard 2007); cumulatively, several studies have shown that striated muscle is an essential source of progenitor cells capable of osteogenic differentiation (Usas and Huard 2007).

After the provocative report that MSCs are the perivascular pericytic stem cells (Crisan *et al.*, 2008), the understanding of several potential biological functions *in vivo* is now possible, including vascular and perivascular driven bone induction in both heterotopic and orthotopic sites (Caplan 2008). Importantly, pericytic cells express and secrete modulators and inducers that contribute important trophic activities by structuring a plastic regenerative microenvironment as initiated within perivascular 'niches' of the striated muscle (Caplan 2008).

The fundamental tenet of the induction of bone formation is to combine osteogenic soluble molecular signals with insoluble signals or substrata to erect scaffolds of biomimetic biomaterial matrices that mimic the supramolecular assembly of the extracellular matrix of bone (Sampath and Reddi 1981; Ripamonti and Reddi 1995; Reddi 2000; Ripamonti *et al.*, 2004). Critical by now studies on the induction of bone formation were performed by implanting allogeneic bone matrices and other matrices in heterotopic sites of rodents and lagomorphs showing unexpectedly the induction of bone formation even when implanted in heterotopic extraskeletal sites of recipient animals (Senn 1889; Sacerdotti and Frattin 1901; Levander 1938; Levander 1945; Levander and Willestaedt 1946; Bridges and Pritchard 1958). Key experimental observations unequivocally reported that demineralized bone matrices when implanted in the extraskeletal heterotopic sites of rodents induce "*bone: formation by autoinduction*" (Urist 1965; Reddi and Huggins 1972). The induction of bone formation by implanting alcohol-extracted and/or demineralized bone matrices prompted a concerted effort to isolate and identify a class or sub-classes of soluble molecular signals endowed with the striking prerogative of inducing *de novo* induction of bone formation (Ripamonti *et al.*, 2004; Ripamonti 2006; Ripamonti *et al.*, 2006; Ripamonti *et al.*, 2008).

Twentieth century research has shown that the intact but demineralized bone matrix retains morphogens (Urist 1965; Reddi and Huggins 1972), firstly defined by Turing as "forms generating substances" (Turing 1952). Morphogens, when released, are capable of imparting variable differentiating pathways to responding stem cells within selected microenvironments. We now know that the extracellular matrix of bone is a multifactorial repository of locally active pleiotropic morphogens that initiate, maintain and modulate in a paracrine and autocrine fashion the cascade of bone differentiation by induction (Reddi 1984; Reddi 2000; Ripamonti 2006; Ripamonti *et al.*, 2006). Which are the molecular signals that initiate the cascade of bone differentiation by induction? Or perhaps, more importantly, where are the soluble molecular signals that orchestrate the bone induction cascade secreted, expressed and stored?

The fundamental work of Sacerdotti and Frattin, Huggins, Levander, Lacroix, Moss, Trueta, Urist, Reddi and others (reviewed by Ripamonti *et al.*, 2006 and Ripamonti *et al.*, 2008) provided evidence for the existence of a bone morphogenetic complex within the bone matrix as well as other extracellular matrices responsible for "*the bone induction principle*" (Urist *et al.*, 1967; Urist *et al.*, 1968; Reddi and Huggins 1972). Where does this "osteogenic activity" of the bone matrix reside? Until recently it was still unclear whether the bone forming activity or osteogenic activity of the intact demineralized bone matrix was due to the combined action of several morphogenetic factors known to be present within the extracellular matrix of bone, a separate protein, or a family of proteins as yet to be characterized and sequenced (Ripamonti and Reddi 1995; Ripamonti *et al.*, 2004; Ripamonti 2006; Ripamonti *et al.*, 2006).

Identification of several osteogenic proteins present within the bone matrix has been hindered by the realization that the bone matrix is in the solid state (Reddi 1997). A solubilised bone morphogenetic protein complex was first obtained by Urist *et al* (1979) who sequentially extracted intact demineralized bone matrix in Hanks' solution containing 300mM NaCl, 3mM $NaN_3$, 25mM Tris with collagenase (Urist *et al.*, 1979).The morphological and biochemical problem of the "*bone matrix in the solid state*" (Reddi 1997) was further compounded by the small quantities of proteins tightly bound to the organic and inorganic components of the extracellular matrix of bone (Sampath and Reddi 1984).

The morphological and biochemical impasse of the "*bone matrix in the solid state*" (Reddi 1997) was unlocked by solubilizing the putative osteogenic proteins from the extracellular matrix of bone (Sampath and Reddi 1981; Sampath and Reddi 1983; Sampath and Reddi 1984). A fundamental step forward, which our laboratories believe set the emergence of the tissue engineering paradigm, i.e. reconstituting soluble molecular signals with extracellular matrix substrata, was the classic work of Reddi and co-authors who unlocked the morphological and biochemical problem of the "*bone matrix in the solid state*" (Reddi 1997) by solubilising extracellular matrix proteins from the bone matrix (Sampath and Reddi 1981; Sampath and Reddi 1983; Ripamonti and Reddi 1995).

The realization that the intact and demineralized bone matrix could be dissociatively extracted and inactivated with chaotropic agents such as 6M guanidinium hydrochloride or urea (Sampath and Reddi 1981) has shown that the bone matrix is a reservoir of multiple molecular signals.

The realization that bone matrix is a reservoir of multiple morphogenetic signals has vindicated Urist's theory of a hypothesized though rationalized bone morphogenetic protein complex within the bone matrix (Urist et al. 1968; Urist and Strates 1971; Sampath and Reddi 1981; Sampath and Reddi 1983). Importantly, the osteogenic activity of the intact demineralized bone matrix, lost after dissociative extraction of the bone matrix, could be re-activated and restored by reconstituting or recombining the extracted inactive and insoluble collagenous bone matrix with the solubilised proteinaceous component (Sampath and Reddi 1981; Sampath and Reddi 1983; Ripamonti and Reddi 1995). The reconstitution resulted in the induction of bone formation after the proteinaceous extract was partially purified by gel filtration chromatography to remove high molecular weight contaminants (Sampath and Reddi 1981).

The operational reconstitution of solubilised osteogenic molecular signals with an insoluble signal or substratum was a key experiment that provided the development of 1), an extraskeletal heterotopic bioassay for *bona fide* initiators of endochondral bone induction, 2), the development and application of increasingly refined purification schemes mainly involving liquid chromatography on the solubilised protein fractions (Wang *et al.*, 1988; Luyten *et al.*, 1989; Sampath *et al.*, 1992; Ripamonti *et al.*, 1992) that 3), resulted in the identification, purification and isolation of an entirely new class of proteins initiators, the bone morphogenetic/osteogenic proteins (BMPs/OPs), powerful inducers of bone formation (Reddi 2000; Ripamonti *et al.*, 2004; Ripamonti 2006; Ripamonti *et al.*, 2006).

Importantly, Reddi's laboratories, then at the National Institutes of Health, Bone Cell Biology Section, Bethesda US, further demonstrated that osteogenic proteins, extracted and partially purified from bone matrices of different animal models reproducibly initiate the cascade of bone differentiation in the rodent subcutaneous bioassay, providing that the solubilised osteogenic proteins are reconstituted with the recipient rat allogeneic insoluble collagenous bone matrix (Sampath and Reddi 1983).

The above studies indicated that there is homology between bone inductive proteins from human, monkeys, bovine and rat bone extracellular matrices (Sampath and Reddi 1983). The insoluble signal, the inactive and insoluble extracted collagenous matrix, thus retains the alloantigenic load, and the initiation of bone formation is only triggered when allogeneic, but not xenogeneic collagenous bone matrices are reconstituted and implanted in heterotopic sites (Sampath and Reddi 1983).

The homology of the isolated osteogenic proteins has been shown by the purification of large quantities of bovine and baboon bone matrices as a starting point for the purification of osteogenic proteins with biological activity in the rodent subcutaneous bioassay only when reconstituted with allogeneic collagenous matrices as carrier (Wang *et al.*, 1988; Luyten *et al.*, 1989; Ripamonti *et al.*, 1992). Of note, highly purified naturally-derived BMPs/OPs from bovine bone matrices induce periodontal tissue regeneration when implanted in furcation osseous defects of the non-human primate *Papio ursinus* (Ripamonti *et al.*, 1994) (Figure 1). Purification to homogeneity resulted in the isolation, identification and cloning of an entirely new family of protein initiators, collectively called BMPs/OPs, members of the transforming growth factor-β (TGF-β) supergene family (Wozney *et al.*, 1988; Wang *et al.* 1990).

Molecular cloning of the now available recombinant human proteins, hBMP-2 and hBMP-7, the latter also known as osteogenic protein-1 (hOP-1), has resulted in extensive testing in pre-clinical settings including non-human primates (Ripamonti *et al.*, 1996; Ripamonti *et al.*, 2000) as well as human primates in clinical contexts (Friedlander *et al.*, 2001; Govender *et al.*, 2002; Gautshi *et al.*, 2007; Garrison *et al.*, 2007). Two key chromatographic steps were fundamental for purification of naturally-derived BMPs/OPs from bovine (Wang *et al.*, 1988; Luyten *et al.*, 1989) and baboon (Ripamonti *et al.*, 1992) bone matrices.

Quantities of intact demineralized bone matrix are demineralized in hydrochloric acid pH 0.01 in three volume changes; proteins are then solubilised in chaotropic agents (Sampath and Reddi 1981; Sampath and Reddi 1983).

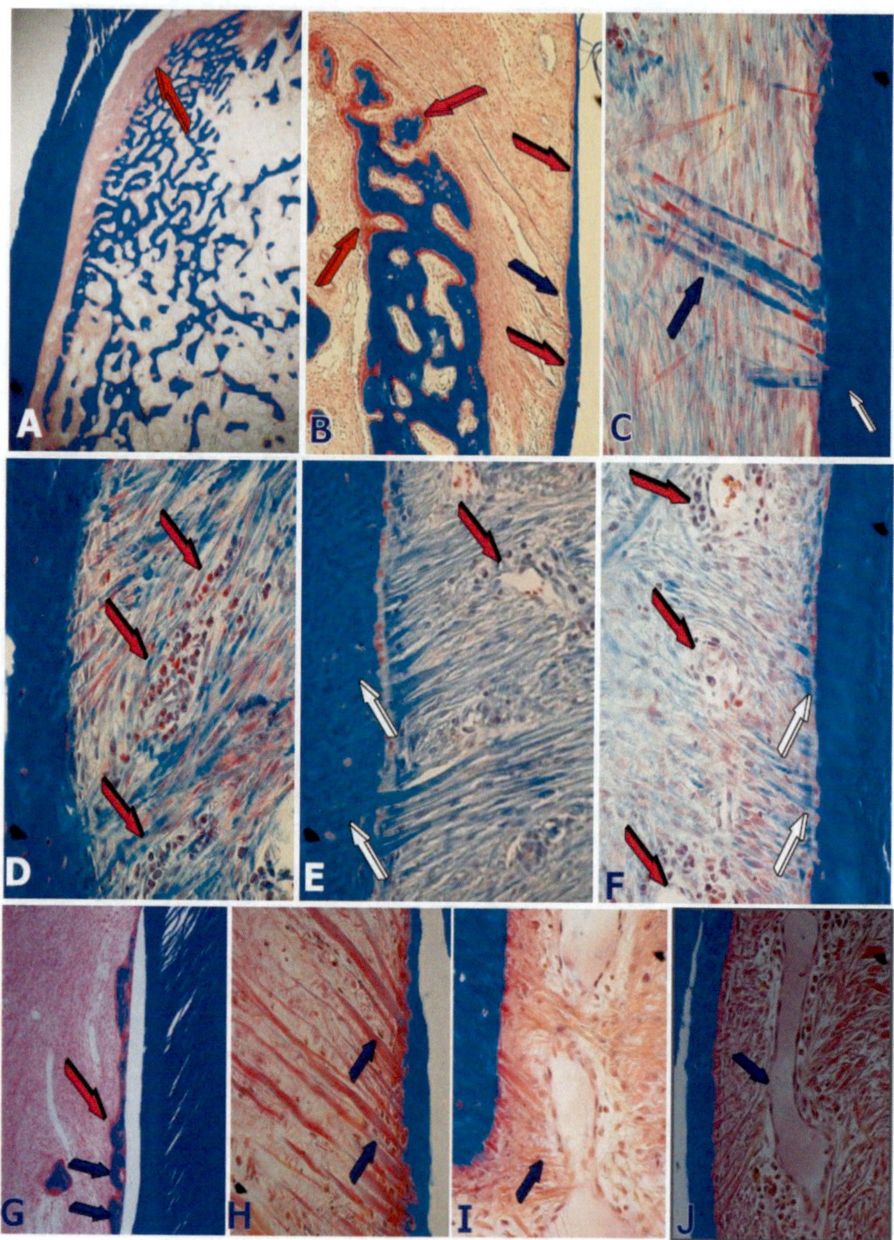


Figure 1. Multiple pleiotropic activities of highly purified naturally-derived osteogenic proteins extracted and purified from bovine bone matrices (Luyten *et al.*, 1987) after adsorption and affinity chromatography on hydroxyapatite-Ultrogel and Heparin-Sepharose columns; protein fractions in the highly osteogenic 500mM NaCl Heparine-

Sepharose eluate were subjected to molecular sieve gel filtration on tandem S-200 Sephacryl chromatography columns (Luyten *et al.*, 1987; Ripamonti *et al.*, 1992). 250μg of highly purified osteogenic proteins after molecular sieve gel filtration with an apparent molecular mass range of 26-42kDa purified greater than 70,000 fold were combined with 150mg of allogeneic inactive insoluble collagenous bone matrix and implanted in Class II furcation defects of non-human primates *Papio ursinus* (Ripamonti *et al.*, 1994). (A): Low power view of a treated furcation defect 60 days after healing showing alveolar bone regeneration with newly formed mineralized bone in blue covered by osteoid seams (*red* arrows) in A and B; (B): Induction of cementogenesis (***blue*** arrow) along the exposed root surface with newly formed mineralized cementum (***blue*** arrow) covered by a thin layer of cementoid as yet to be mineralized newly formed cemental matrix (*magenta* arrows) populated by cementoblasts; (C): high power view of *de novo* generation of Sharpey's fibers into dentine matrix surfaced by a thin layer of newly induced cementoid with secreting cementoblasts interspersed within the newly formed fibers (blue arrow); note how the fibers tightly insert deep into mineralized dentine matrix (white arrow). Highly purified osteogenic proteins are not only osteogenic restoring the induction of bone formation in the implanted surgically-created furcation defects (A) but also cementogenic as shown in B and C, with generation of newly formed cementoid matrix covered by cementoblasts (B); osteogenic proteins are also inducers of bona fide mineralized Sharpey's fibers that directly insert into the mineralized dentine (*white* arrow in C); (D,E,F): the vast and multiform pleiotropic activities are additionally shown by the induction of angiogenesis (magenta arrows) in D,E,F; (D): *magenta* arrows point to cellular elements with condensed chromatin indicating de novo angiogenesis within the regenerated periodontal ligament space; (E,F): Sharpey's fibers penetrating the dentine (*white* arrows) between cementoblasts actively secreting cementoid matrix at the dentine interface; magenta arrows indicate capillary sprouting within the newly formed and assembled periodontal ligament space; (G): Newly formed mineralized cementum in (blue arrows) with remnants of cementoid matrix as yet to be mineralized (*magenta* arrow) with inserted newly generated Sharpey's fibers; (H): detail of newly formed and mineralized cementum surfaced by cementoid matrix with multiple cementoblasts showing cellular trafficking at the cemental interface with elongated fibroblast-like cells riding single collagenic fibers (***blue*** arrows); (I,J): Intimate and exquisite morphological and thus molecular relationships between newly formed sprouting capillaries, synthesized periodontal ligament fibers and 'riding' osteoblast and/or cementoblast progenitors depending on the site/specific morphogen gradients of the periodontal ligament space. (I,J): ***blue*** arrows indicate the morphologically exquisite merging of periodontal ligament fibers within the extracellular matrix component of the newly formed capillaries providing the supramolecular and cellular assembly of the periodontal ligament space; (J): Pending on selected morphogenetic gradients across the periodontal ligament space, cementoblast and/or osteoblast progenitors migrate out of the endothelial perivascular stem cell 'niche', encroach single collagenic fibers and ride the fibers at its phenotypic end constructing either cementum as cementoblasts or bone as osteoblasts. Undecalcified sections cut at 3μm stained free floating with Goldner's trichrome; Undecalcified sections courtesy of Barbara van den Heever, Bone Research Laboratory.

The first preparative adsorption chromatography is on hydroxyapatite-Ultrogel chromatography to which BMPs/OPs greatly bind later eluted with column buffer of 100mM phosphate (Luyten *et al.*, 1989; Sampath *et al.*, 1992; Ripamonti *et al.*, 1992). The eluted 100mM phosphate pick, concentrated and exchanged to 6M urea pH 7.4 is then loaded onto a heparin-Sepharose affinity chromatography column. The 500mM NaCl eluate, concentrated and exchanged to 6M guanidinium hydrochloride is then loaded onto in tandem Sephacryl S-200 gel filtration chromatography columns and eluted fractions bioassayed after implantation in the subcutaneous space of the rat (Luyten *et al.*, 1989; Sampath *et al.*, 1992; Ripamonti *et al.*, 1992).

Protein fractions with an apparent molecular weight of 30-42KDa retain the highest biological activity when implanted in the subcutaneous space of the rat (Luyten *et al.*, 1989; Ripamonti *et al.* 1992); final purification to homogeneity is obtained by electroendosmotic elution from a preparative sodium dodecyl sulphate (SDS) polyacrylamide gel, resulting in a single band on a SDS-polyacrylamide gel with an apparent molecular mass of 30-34kDa (Figure 2), with biological activity in rats (Luyten *et al.*, 1989; Ripamonti *et al.*, 1992). Highly purified protein fractions can then be used for implantation in non-human (Ripamonti *et al.*, 1992; Ripamonti *et al.*, 1993; Ripamonti *et al.,* 2001) and human (Ripamonti and Ferretti 2002; Ferretti and Ripamonti 2002) primates.

The capacity of mammalian BMPs/OPs to initiate a programmed cellular cascade that results in the induction of bone is a functionally conserved process utilized in embryonic development and recapitulated in postnatal osteogenesis that can be exploited for the therapeutic initiation of bone formation (Ripamonti *et al.*, 1992).

Previous studies in the non-human primate *Papio ursinus* have shown that recombinant human osteogenic protein-1 (hOP-1) is capable of inducing complete regeneration of large cranial defects in adult primates, demonstrating the therapeutic utility of a single application of hOP-1 in preclinical contexts (Ripamonti *et al.*, 1996). Complete regeneration was observed by day 90, with large trabeculae interspersed with marrow uniting the pericranial and the endocranial cortices, characterized by the presence of remodelled bone (Ripamonti *et al.*, 1996).

Continuous osteogenesis was indicated by the presence of osteoid seams populated by osteoblasts, followed by remodelling of the newly formed osteonic lamellar bone by day 365 (Figure 3) (Ripamonti *et al.*, 1996). Importantly, the observation of exuberant osteogenesis on the pericranial sites of the defect suggested that resident osteoprogenitor cells in the pericranium (periosteum) and myoblastic cells within the overlaying *temporalis* muscle are the primary target for hOP-1 in preclinical settings (Ripamonti *et al.*, 1996; Ripamonti *et al.*, 2000).

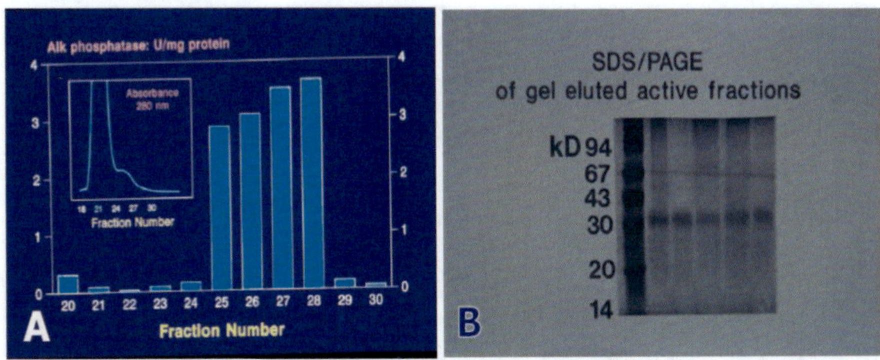


Figure 2. Purification, biological activity and electrophoretic profile of osteogenin, a bone morphogenetic protein, extracted and purified greater than 70,000 fold from baboon bone matrices (Ripamonti *et al.*, 1992). (A): Alkaline phosphatase activity at day 11 of implants of rat insoluble collagenous bone matrix reconstituted with baboon osteogenin fractions after Sephacryl S-200, bioassayed in the subcutaneous space of Long-Evans rats. The osteogenic activity was confined solely to fractions with an apparent molecular mass of 26–42kDa. *Inset*: Sephacryl S-200 gel filtration profile. Protein fractions with the highest biological activity as determined by the heterotopic bioassay in rodents are isolated in a single shoulder after gel filtration chromatography (Ripamonti *et al.*, 1992). (B): Electrophoretic profile of osteogenin on SDS-polyacrylamide gel under non-reducing conditions of electroendosmotic eluted osteogenic fractions after preparative SDS gel electrophoresis of the bioactive fractions after S-200 gel filtration chromatography. To estimate protein concentration at the nanogram level, protein bands were silver stained, and the gel was scanned at 580nm to estimate protein content in reference to known amounts of native bovine osteogenin electroeluted and electrophoresed as described for baboon protein fractions (Luyten *et al.*, 1989). (B): Molecular mass markers are given in kDa. Purification to homogeneity courtesy of Laura Yeates, 1991 Dental Research Institute and National Institutes of Health, Bone Cell Biology Section (Ripamonti *et al.*, 1992).

The application of the osteogenic soluble signal (hOP-1) reconstituted with an insoluble signal or substratum (the insoluble and inactive collagenous bone matrix) for the induction of tissue morphogenesis and regeneration illustrates the importance of the extracellular matrix for cell recruitment, attachment, proliferation and differentiation (Sampath and Reddi 1981; Reddi 1984; Reddi 2000). Of note, the long-term study of hOP-1 implanted in calvarial defects of *Papio ursinus* deployed both allogeneic baboon and xenogeneic bovine insoluble collagenous bone matrices as carriers (Ripamonti *et al.*, 1996).

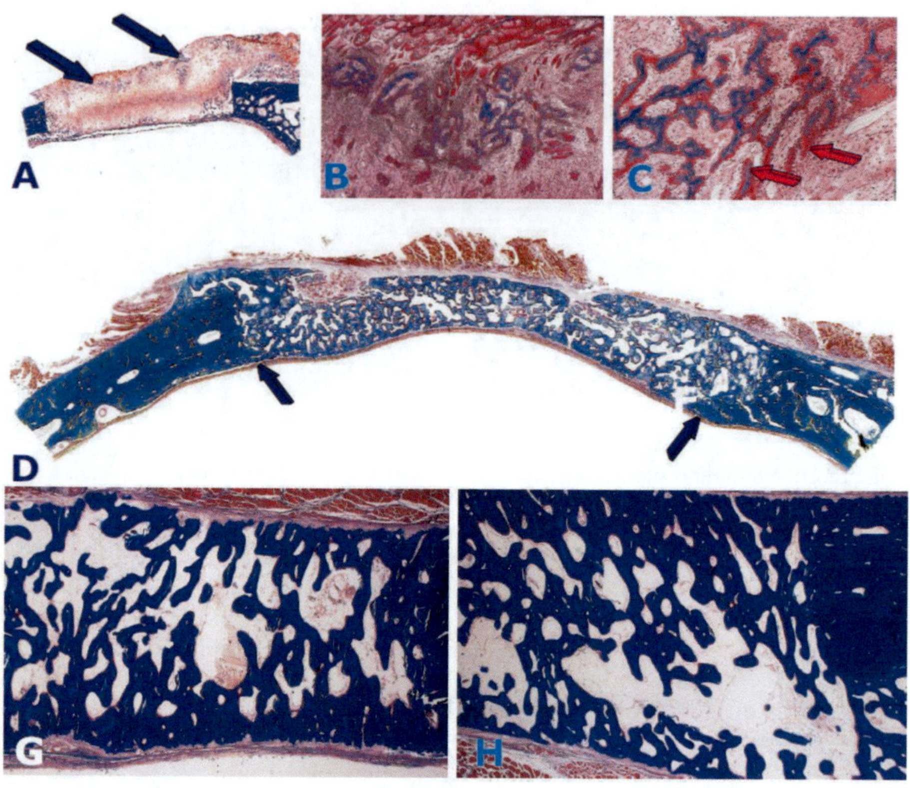


Figure 3. Morphology of calvarial regeneration and induction of bone formation in membranous bone defects surgically prepared in the calvaria of non-human primates *Papio ursinus* after implantation of gamma-irradiated hOP-1 osteogenic devices harvested on day 15 (A, B, C) and 90 (D,E,F); (A): Pericranial and endocranial osteogenesis (***blue*** arrows) by 0.1mg gamma-irradiated hOP-1; (B,C): High power views of pericranial osteogenesis by the hOP-1 osteogenic device with newly formed bone in ***blue*** surfaced by continuous osteoid seams (***magenta*** arrows) in B,C; (D): complete regeneration of the calvarial defect (***blue*** arrows) 90 days after implantation of the gamma-irradiated hOP-1 osteogenic device; (G,H): Solid blocks of remodeled bone 90 days after implantation of the 2.5mg gamma-irradiated hOP-1 osteogenic device showing reconstruction and maintenance of the calvarial profile across the defect. Undecalcified sections cut at 5μm stained free floating with Goldner's trichrome; Undecalcified sections courtesy of Barbara van den Heever, Bone Research Laboratory.

Interestingly, insoluble matrix derived from baboon or bovine sources had a different effect on the rate of tissue induction and remodelling when generated tissue areas were analyzed (Ripamonti *et al.*, 1996). The results obtained with the bovine collagenous matrix as a carrier additionally suggested that allogeneic matrices may not be a requirement for the therapeutic application of a recombinant protein, in context, the hOP-1 osteogenic device (Ripamonti *et al.*, 1996).

The above studies were fundamental to use in pre-clinical contexts xenogeneic bovine bone matrices as a carrier for the biological activity of the hOP-1, devising the hOP-1 osteogenic device for implantation in clinical contexts (Ripamonti *et al.*, 1996). Further studies were thus implemented in *Papio ursinus* to investigate the long-term efficacy of gamma-irradiated hOP-1 in bone tissue induction and regeneration combined with a bovine collagenous matrix as carrier, sterilized with 2.5Mrads of gamma-irradiation, and implanted in 80 calvarial defects of 20 adult Chacma baboon *Papio ursinus* (Figure 3) (Ripamonti *et al.*, 2000; Ripamonti 2005). One year after the implantation of the irradiated hOP-1 devices, bone and osteoid volumes and generated bone tissue areas were comparable with non-irradiated hOP-1 (Ripamonti *et al.*, 1996; Ripamonti *et al.*, 2000; Ripamonti 2005). Interestingly, 365 days after implantation, regenerates induced by 0.5 and 2.5 mg of gamma-irradiated hOP-1 devices showed greater amounts of bone and osteoid volumes when compared with results obtained by non-irradiated hOP-1 devices (Ripamonti *et al.*, 1996; Ripamonti *et al.*, 2000; Ripamonti 2005). Of note, control specimens of gamma-irradiated collagenous matrix without hOP-1 displayed a nearly two-fold reduction in osteoconductive bone repair when compared with non-irradiated controls (Ripamonti *et al.*, 2000). The reduction in bone volume and bone tissue area is caused by a reduced performance of the gamma-irradiated collagenous bone matrix substratum rather than to a reduction of the biological activity of the irradiated recombinant hOP-1 (Ripamonti *et al.*, 2000).

The above conclusions are supported by the results of *in vitro* and *in vivo* studies performed to determine the structural integrity of the recovered gamma-irradiated hOP-1 prior to implantation (Ripamonti *et al.*, 2000). Recoveries by high performance liquid chromatography and sodium dodecylsulphate/polyacrylamide gel electrophoresis (SDS/PAGE) immune blot analyses indicated that doses of 2.5-3Mrads of gamma-irradiation did not significantly affect the structural integrity of the recombinant hOP-1; importantly, biological activity of the recovered hOP-1 was confirmed *in vitro* by the induction of alkaline phosphatase activity in rat osteosarcoma cells and *in vivo* by *de novo* endochondral bone formation in the subcutaneous space of the rodent bioassay (Ripamonti *et al.*, 2000).

Additional studies using gamma-irradiated hOP-1 osteogenic devices were designed to study the incorporation of gamma-irradiated bovine collagenous bone matrices recombined with gamma-irradiated hOP-1 implanted in both extraskeletal heterotopic and calvarial orthotopic sites of adult non-human primates *Papio ursinus* (Figure 3) (Ripamonti 2005). Predictable bone induction in clinical contexts requires information on the expression and cross regulation of gene products of the TGF-β superfamily elicited by single applications of each recombinant hBMPs/OPs. Using the calvarium and the *rectus abdominis* muscle of *Papio ursinus* as a model for tissue induction and morphogenesis, the study investigated the induction of bone formation by gamma-irradiated hOP-1 delivered by gamma-irradiated bovine insoluble collagenous bone matrices for bone induction in heterotopic and orthotopic sites in *Papio ursinus* (Ripamonti 2005). Of note, the expression patterns of OP-1, collagen type IV, BMP-3 and TGF-β mRNAs elicited by increasing single applications of doses of the hOP-1 osteogenic device were also studied after implantation of 0.1, 0.5 and 2.5 mg hOP-1 per gram of gamma-irradiated bovine collagenous matrix in the *rectus abdominis* muscle and in orthotopic calvarial defects of 12 adult Chacma baboons *Papio ursinus* (Ripamonti 2005). Histology and histomorphometry on serial undecalcified sections prepared from the specimens harvested on day 15, 30 and 90 showed that all the doses of the hOP-1 osteogenic devices induced bone formation culminating in complete calvarial regeneration by day 90 (Figure 6). Type IV collagen mRNA expression, a marker of angiogenesis, was strongly expressed in both heterotopic and orthotopic tissues. High levels of expression of OP-1 mRNA demonstrated autoinduction of OP-1 mRNAs. Expression of BMP-3 mRNA varied from tissues induced in heterotopic *vs.* orthotopic sites with high expression in rapidly forming heterotopic ossicles together with high expression of type IV collagen mRNA (Ripamonti 2005). The temporal and spatial expressions of TGF-$\beta_1$ mRNA indicated a specific temporal transcriptional window during which expression of TGF-$\beta_1$ is mandatory for successful and optimal osteogenesis (Ripamonti 2005). Importantly, the study concluded that the induction of bone formation by a single recombinant human protein, the OP-1 osteogenic device, develops as a mosaic structure with distinct spatial and temporal patterns of gene expression of members of the TGF-β supergene family that singly, synergistically and synchronously initiate and maintain tissue induction and morphogenesis (Ripamonti 2005).

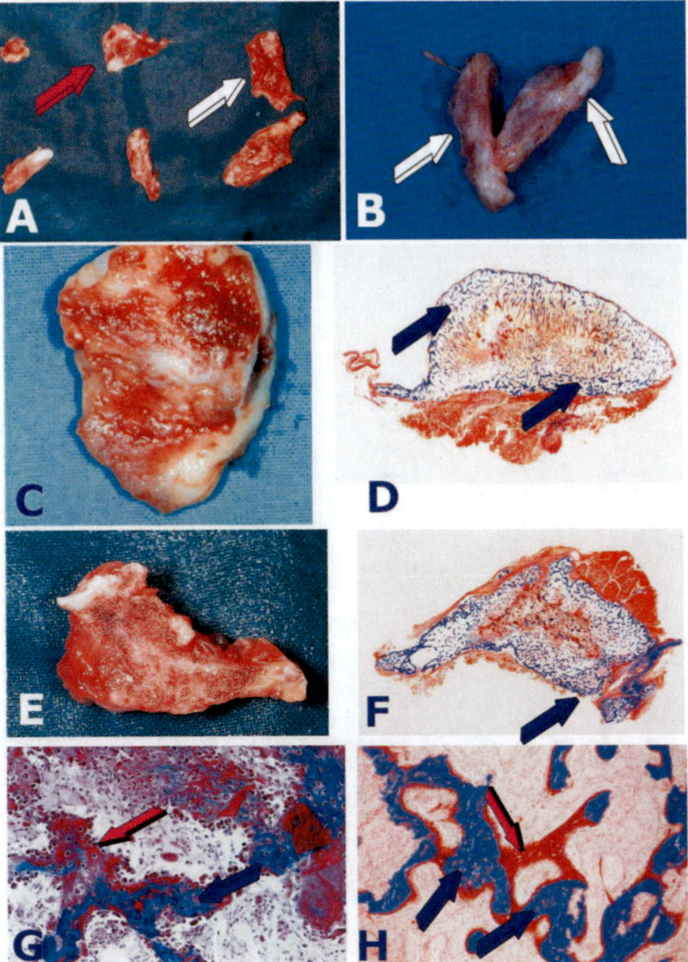


Figure 4. Prominent and robust induction of bone formation by doses of the recombinant human transforming growth factor-$\beta_3$ (hTGF-$\beta_3$) reconstituted with allogeneic insoluble collagenous bone matrix to form the osteogenic device implanted in heterotopic intramuscular sites of the *rectus abdominis* muscle of the non-human primate *Papio ursinus*, and harvested on day 30. (A,B): Large ossicles after implantation of 25 (***magenta*** arrow) and 125 (*white* arrows) μg hTGF-$\beta_3$ reconstituted with allogeneic baboon insoluble collagenous bone matrix as carrier; (C,D,E,F): Large corticalized ossicles harvested from the *rectus abdominis* muscle 30 days after implantation; (D,F): Corticalization (***blue*** arrows) of the newly formed mineralized bone enveloping scattered remnants of collagenous matrix as carrier; (G,H): High power views of newly generated mineralized bone (***blue*** arrows) surfaced by osteoid seams (***magenta*** arrows) populated by contiguous osteoblasts. Undecalcified sections cut at 5μm stained free floating with Goldner's trichrome.

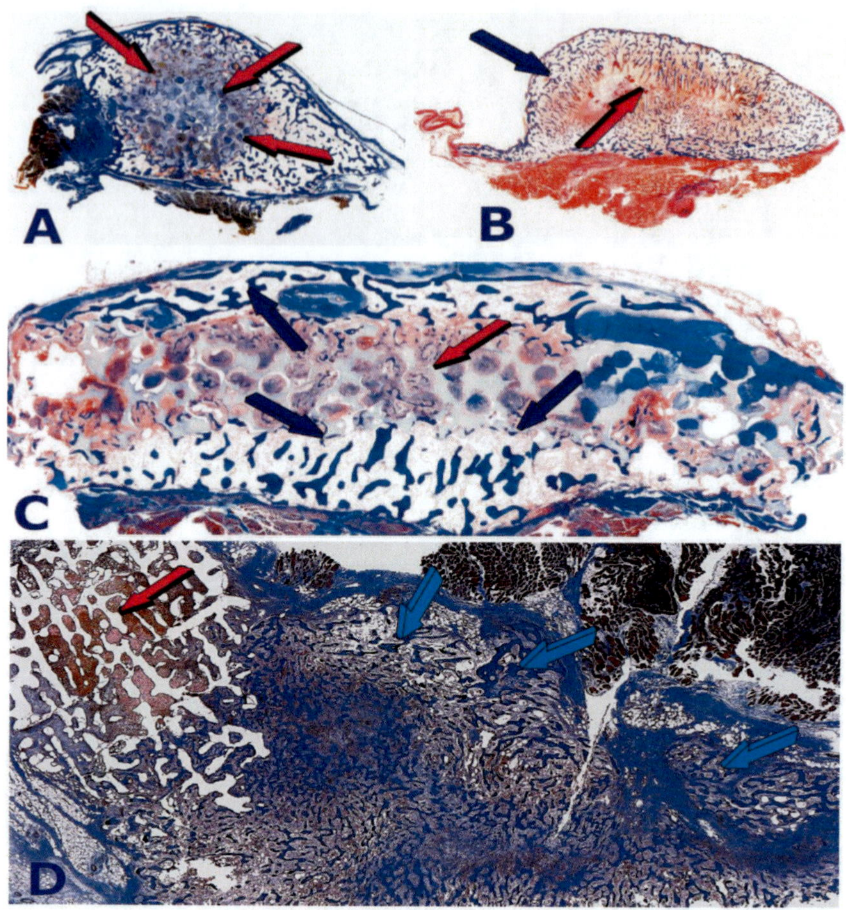


Figure 5. Rapid heterotopic induction of bone formation by the recombinant human transforming growth factor-$\beta_3$ (hTGF-$\beta_3$) reconstituted with macroporous hydroxyapatite-based biomimetic matrices implanted in the *rectus abdominis* muscle of *Papio ursinus*. (A,B): Almost similar tissue induction, morphogenesis and spatial relationship within the *rectus abdominis* muscle when biphasic hydroxyapatite/β-tricalcium phosphate (A) or insoluble collagenous bone matrices (B) are used to deliver the biological activity of the hTGF-$\beta_3$ isoform; (C): prominent induction by a disc of macroporous calcium-phosphate carrier (***magenta*** arrow) with significant osteoinduction outside (***blue*** arrows) the periphery of the implanted macroporous matrix also outlined by magenta arrows in A; (D): Substantial and as yet unreported osteogenesis by 250μg hTGF-$\beta_3$ adsorbed onto a coral-derived calcium phosphate/calcium carbonate macroporous device (***magenta*** arrow); Note the prominent extension of the newly formed bone (***blue*** arrows) well outside the perimeter of the implanted coral-derived macroporous matrix. Decalcified sections cut at 5μm.

Bone tissue engineering in clinical contexts, however, has proven to be an elusive target when compared to results obtained in pre-clinical studies including non-human primates (Ripamonti *et al.*, 2006; Ripamonti *et al.*, 2007; Ripamonti *et al.*, 2012). Tissue engineering of bone in clinical contexts is the culmination of several decades of concerted research on regenerative medicine of the axial and craniofacial skeletons.

This edifice was founded on the results of an extraordinary volume of animal research including non-human primates (Ripamonti 2006; Ripamonti *et al.*, 2012). Preclinical studies particularly in non-human primates have indicated that recombinant hBMPs/OPs together with other members of the TGF-β supergene family were endowed with the striking prerogative to induce bone formation in mammalian tissue (Reddi 2000; Ripamonti 2006). Several tens of milligrams of a single recombinant BMP/OP are needed to often induce uninspiring bone volumes in human patients (Friedlander *et al.*, 2001; Govender *et al.*, 2002; Ripamonti *et al.*, 2007; Gautschi *et al.*, 2007; Garrison *et al.*, 2007; Ripamonti *et al.*, 2012). The induction of bone formation has dramatically shown that regenerative medicine in clinical contexts is on a different scale altogether when compared to animal models that may not adequately translate and reproduce morphogen-related therapeutic responses in *Homo sapiens* (Ripamonti 2010). Off-label use of hBMP-2 and hOP-1 in the maxillofacial skeleton was initially limited to single case reports with enthusiastic conclusions to uninspiring results. The enthusiasm was again based on the histological evidence of osteoinduction. Clinical assessment routinely revealed weak evidence of bone regeneration (Ripamonti *et al.*, 2006; Ripamonti *et al.*, 2007; Ripamonti *et al.*, 2012). Subsequent reports in the maxillofacial region have testified to the lack of clinical performance comparable to autogenous bone grafts. These failures have often been dismissed, and many authors continue to endorse single hBMP-based therapeutic strategies.

The recent revelations of significant complications and failure of osteoinduction in spinal application should sound the death knell for the current philosophy of a single recombinant morphogen at an inflexible dose with the same delivery system (Ripamonti *et al.*, 2012).

Moreover, the acid test for clinically relevant bone tissue engineering should now become the concept of *clinically significant osteoinduction* i.e. the regenerated bone must be readily identifiable on radiographic examination as bone by virtue of its opacity and its trabecular structure. Reliance of histology as a measure of success in clinical settings should be diminished if not discontinued. The need to explore alternative avenues including other osteogenic members of the TGF-β supergene family either solo or in binary application with myoblastic paravascular stem cells with more rigorous criteria for success is now more acutely felt than ever before (Ripamonti *et al.*, 2012).

In contrast to studies in rodents and lagomorphs (Roberts *et al.*, 1986), heterotopic implantation of human recombinant transforming growth factor-$\beta_3$ (hTGF-$\beta_3$) in the non-human primate *Papio ursinus* induces substantial bone formation (Figures 4, 5) (Ripamonti *et al.*, 2008; Ripamonti *et al.*, 2012). Doses of 125μg hTGF-$\beta_3$ induce significant osteogenesis in full thickness mandibular defects of *Papio ursinus* with unprecedented *restitutio ad integrum* as early as 30 days (Ripamonti, 2006a). At the same time, implantation of hTGF-$\beta_3$ in calvarial defects induces over expression of Smad-6 and -7 inhibiting the bone induction cascade (Ripamonti *et al.*, 2008).

The substantial induction of bone formation by day 30 in non-healing mandibular defects in *Papio ursinus* (Ripamonti, 2006a) prompted experimental surgical reconstruction in human subjects (Ripamonti and Ferretti 2012); 125μg and 250μg hTGF-$\beta_3$ per gram of human demineralized bone matrix were implanted in two pediatric patients respectively, 10 to 14 grams per mandibular defect after surgical removal of the hemi-mandibles (Ripamonti and Ferretti 2012; Ripamonti *et al.*, 2012). Radiographic analyses of the reconstructed mandibles show the induction of bone formation across the defects.

In concurrent studies, 250μg hTGF-$\beta_3$ recombined with calcium phosphate-based macroporous constructs resulted in massive induction of bone formation well outside the profile of the implanted scaffold as early as 20 days after heterotopic implantation in *Papio ursinus* (Figure 5) (Ripamonti *et al.*, 2012). The rapid induction of bone formation by hTGF-$\beta_3$ together with *TGF-$\beta_1$*, *BMP-3* and *OP-1* mRNA expression, hypercellular osteoblastic activity, osteoid synthesis, angiogenesis and capillary sprouting have suggested the novel molecular and morphological basis for the induction of bone formation in clinical contexts (Ripamonti *et al.*, 2008; Ripamonti and Ferretti 2012; Ripamonti *et al.*, 2012). Indeed, our last treated pediatric patient was implanted with the 250μg of the hTGF-$\beta_3$ osteogenic device, translating research data from pre-clinical results in non-human primates *Papio ursinus* to clinical contexts (Ripamonti and Ferretti 2012).

The need for alternatives to recombinant human bone morphogenetic proteins is now felt more acutely after reported complications and performance failure in clinical applications (Ripamonti *et al.*, 2012). In the *bona fide* heterotopic assay for bone induction in rodents, the three mammalian TGF-β isoforms do not initiate endochondral bone induction.

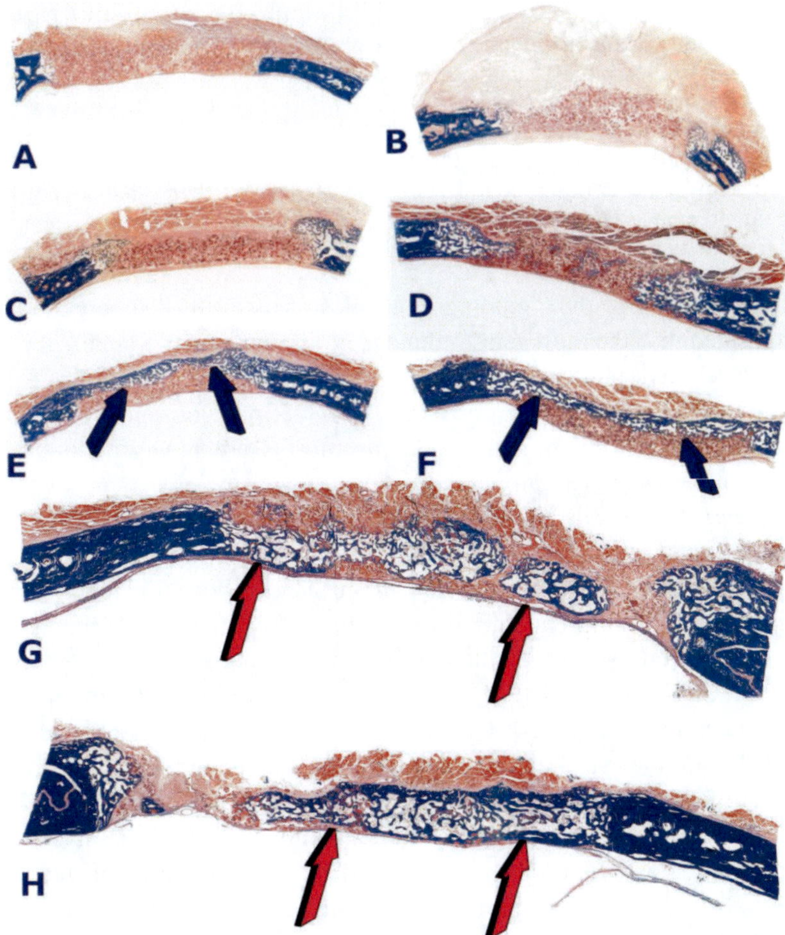


Figure 6. Morphology of calvarial regeneration and induction of tissue morphogenesis by the transforming growth factor-$\beta_3$ (hTGF-$\beta_3$) delivered by allogeneic insoluble collagenous bone matrix with or without recombinant human Ebaf/Lefty-A, a new member of the transforming growth factor-β supergene family (Ripamonti *et al.*, 2005); (A,C): limited if any induction of bone formation by 125μg hTGF-$\beta_3$ recombined with insoluble collagenous bone matrices as carrier; (B,D): osteogenesis, though limited after binary applications of hTGF-$\beta_3$ and hEbaf/Lefty-A delivered by insoluble collagenous bone matrices as carrier. Note in B prominent induction of tissue formation resulting in yet limited calvarial bone formation on day 90 (D). (E,F): On day 90, hTGF-$\beta_3$-treated specimens often showed pericranial induction of bone formation across the defects (***blue*** arrows in E and F); This morphological observation suggested that the *temporalis* muscle is a source of readily available myoblastic muscle-derived stem cells capable of direct transformation into osteoblastic-like cells. The observation has suggested combining the human transforming growth factor-$\beta_3$

(hTGF-$\beta_3$) osteogenic device to morcellated segments of *rectus abdominis* muscle thus providing a readily available number of stem cells for direct transformation into osteoblastic-like cells actively synthesizing newly formed bone across the treated calvarial defects. Indeed, morcellated *rectus abdominis* fragments combined with 125μg doses of the hTGF-$\beta_3$ osteogenic device partially restored the induction activity of the hTGF-$\beta_3$ across the treated defects (G, H) also inducing endocranial bone (***magenta*** arrows), never observed in hTGF-$\beta_3$–treated defects harvested on day 30 or 90 (Ripamonti *et al.*, 2008; Ripamonti *et al.*, 2009). Undecalcified sections cut at 7μm stained free floating with Goldner's trichrome.

The pleiotropy of the signaling molecules of the TGF-β supergene family is highlighted by the apparent redundancy in molecular signals initiating endochondral bone induction but in the primate only (Ripamonti *et al.*, 1997; Ripamonti *et al.*, 2000; Ripamonti *et al.*, 2008; Ripamonti and Roden 2010). Strikingly, the three mammalian TGF-β isoforms are powerful inducers of endochondral bone when implanted in the *rectus abdominis* muscle of the primate *Papio ursinus* at doses of 5, 25, and 125 μg per 100 mg of insoluble and inactive collagenous matrix as carrier, yielding corticalized ossicles by day 90 (Figures 4, 5) and expression of mRNA of bone induction markers (Ripamonti *et al.*, 2000; Ripamonti *et al.*, 2008; Ripamonti and Roden 2010). Ossicles generated by the mammalian TGF-β isoforms express mRNA of OP-1, BMP-3, GDF-10 and TGF-$\beta_1$ in heterotopic constructs (Ripamonti *et al.*, 1997; Ripamonti *et al.*, 2000; Ripamonti *et al.*, 2008; Ripamonti and Roden 2010). Of note, the rapid architectural sculpture of mineralized constructs in the *rectus abdominis* particularly by the hTGF-$\beta_3$ isoform solo or in binary application with hOP-1, a synergistic strategy known to yield massive ossicles in heterotopic sites (Ripamonti *et al.*, 1997), is a novel source of developing auto-induced bone for autogenous transplantation for clinical use (Ripamonti 2010).

The exact mechanisms by which TGF-β signaling results in induction of bone formation in non-human primates *Papio ursinus* still remains to be characterized (Ripamonti and Roden 2010).

Current research does not as yet provide evidence that results in *Papio ursinus* are predictive of biological activity of the mammalian TGF-β isoforms in *Homo sapiens*, although DNA homologies between primates are certainly higher than homologies between rodents and non-human primates. In parallel experiments, we have reported limited induction of bone formation in orthotopic calvarial defects implanted with doses of all three mammalian TGF-β isoforms (Ripamonti *et al.*, 1996; Ripamonti *et al.*, 2000; Ripamonti *et al.*, 2008; Ripamonti and Roden 2010).

Limited induction of bone formation in orthotopic calvarial defects implanted with hTGF-$\beta_3$ osteogenic devices is due to the influence of Smad-6 and -7 downstream antagonists of the TGF-β signaling pathway (Ripamonti *et al.*, 2008). RT-PCR analyses of newly induced ossicles generated by the hTGF-$\beta_3$ isoform have shown robust expression of Smad-6 and -7 in orthotopic calvarial sites with limited expression in heterotopic *rectus abdominis* sites (Ripamonti *et al.*, 2008). Our morphological and molecular studies have suggested that Smad-6 and -7 over expression in hTGF-$\beta_3$–treated calvarial defects may be due to the vascular endothelial tissue of the arachnoids expressing signaling proteins which modulate the expression of the inhibitory Smads in pre-osteoblastic and osteoblastic cell lines, thus controlling the induction of bone formation in the primate calvarium (Ripamonti *et al.*, 2008). Of note, morcellated fragments of autogenous *rectus abdominis* muscle, containing pericytes, myoendothelial, and myoblastic stem cells (Zheng *et al.*, 2007; Kovacic and Boehm 2009; Péault *et al.*, 2007; Chen *et al.*, 2009; Crisan *et al.*, 2008), significantly increase the induction of bone formation in calvarial defects treated with hTGF-$\beta_3$ delivered by collagenous matrix as carrier (Figure 6) (Ripamonti *et al.*, 2008; Ripamonti *et al.*, 2009c; Ripamonti and Roden 2010).

The addition of responding stem cells prepared by finely morcellating fragments of autogenous *rectus abdominis* muscle significantly enhances the induction of periodontal tissue regeneration when combined with hTGF-$\beta_3$ in Matrigel® matrix implanted in Class II and III furcation defects of *Papio ursinus* (Figure 7) (Ripamonti *et al.*, 2009a; Teare *et al.*, 2008). Importantly, myoblastic stem cells prepared by finely mincing fragments of autogenous *rectus abdominis* muscle significantly increase the coronal extent of cementogenesis along surgically exposed root surfaces (Figure 7) (Ripamonti *et al.*, 2009a; Ripamonti *et al.*, 2009b).

This chapter further describes the induction of bone formation in calvarial defects as modulated by the addition of morcellated *rectus abdominis* muscle fragments to enhance tissue induction and morphogenesis as evaluated in the non-human primate *Papio ursinus* as a prerequisite for potential clinical applications in man.

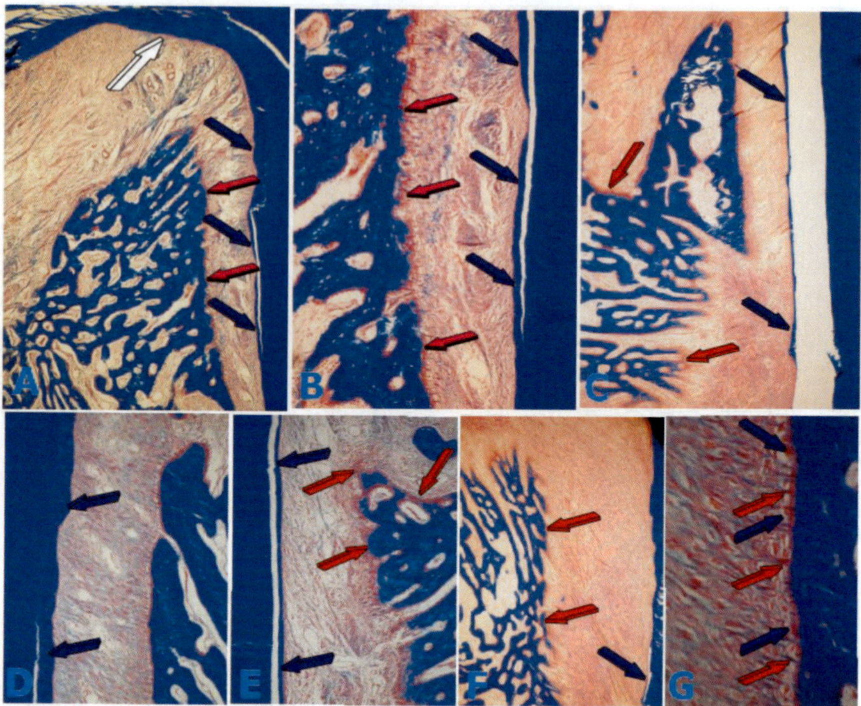


Figure 7. Tissue induction and morphogenesis 75μg hTGF-$\beta_3$-osteogenic device implanted in Class III furcation defects of the non-human primate *Papio ursinus* pre-combined and delivered by 300μl growth factor-reduced Matrigel® matrix together with finely morcellated fragments of autogenous *rectus abdominis* muscle 60 days after implantation (Ripamonti *et al.*, 2009). (A,B,C) Substantial induction of alveolar bone regeneration coronally extending to the furca of the defect (***white*** arrow) together with *de novo* induction of cementogenesis (***blue*** arrows in A and B) along the exposed root surfaces with osteoid seams on the newly formed mineralized bone in ***blue*** with *de novo* generated fibers extending into the periodontal ligament space (***magenta*** arrows in A,B and C); (D,E,F): High power view of the morphology of the regenerated periodontal ligament space, the alveolar bone and cementum by 75μg hTGF-$\beta_3$ in Matrigel® matrix combined with finely morcellated fragments of autogenous *rectus abdominis* muscle (Ripamonti *et al.*, 2009). In all panels, ***blue*** arrows indicate newly formed and mineralized cementum whilst ***red*** arrows indicate newly synthesized as yet to by mineralized osteoid matrix; (G): Multiple *de novo* generated Sharpey's fibers originating within and extruding from the planed dentine matrix (***magenta*** arrows) between cementoblastic cells (***blue*** arrows) actively synthesizing cemental matrix. Undecalcified sections cut at 7μm stained free floating with Goldner's trichrome.

## THE INDUCTION OF BONE FORMATION BY THE MAMMALIAN TGF-B$_3$ ISOFORM

Ever since Hippocrates (400B.C.) reported on the unique phenomenon of bone healing itself with no apparent scarring (Reddi 1994), molecular biologists and tissue engineers alike have attempted to unravel the unique and primary controlling mechanisms by which bone regeneration occurs (Reddi 2000). Although significant strides have been made in the field of hard tissue regeneration, particularly regarding the cytological and structural entities that are required to initiate the bone induction cascade, the molecular mechanistic pathway debate still rages on. Fundamental to our understanding are the key cellular and molecular signals that control the bone induction cascade and how they are applied within the greater biochemical macro- and microenvironments that lead to bone formation by induction.

Prior to 1997, bone morphogenetic proteins/osteogenic proteins (BMPs/OPs), members of the transforming growth factor-beta (TGF-β) superfamily were believed to be the sole initiators of the bone induction cascade (Wozney *et al.*, 1988). When evaluating another peptide subgroup of the TGF-β superfamily, the TGF-β isoforms *per se,* and in particular the mammalian TGF-$\beta_1$, -$\beta_2$, and -$\beta_3$ for its osteogenic activities in rodents, lagomorphs and canines, it was found that the TGF-β isoforms do not induce the cascade of bone differentiation by induction when implanted in heterotopic extraskeletal sites (Roberts *et al.* 1986; Ripamonti 2006). These findings, with respect to the TGF-β isoforms not inducing bone formation, made BMPs appear to be the main regulatory morphogens that induced/stimulated bone development by autoinduction (Urist *et al.*, 1967; Wozney *et al.*, 1988).

In the past several years, the osteogenic activity of selected TGF-β superfamily members has been thoroughly and systematically re-evaluated and tested in heterotopic extraskeletal sites of the non-human primate *Papio ursinus*; the mammalian recombinant hTGF-$\beta_1$, -$\beta_2$ and -$\beta_3$ were implanted in the *rectus abdominis* muscle of the Chacma baboons *Papio ursinus* combined with either insoluble collagenous bone matrix or macroporous calcium phosphate-based biomimetic matrices as carriers (Ripamonti *et al.*, 1997; Ripamonti *et al.*, 2000; Ripamonti *et al.*, 2008; Ripamonti and Roden 2010; Ripamonti *et al.*, 2012). Contrary to previous studies that reported that the mammalian TGF-β isoforms were not inducers of heterotopic endochondral bone (Roberts *et al.*, 1986; Wozney *et al.*, 1988; Ripamonti 2006), the mammalian TGF-β isoforms are powerful inducers of substantial bone formation but only in non-human primates, in particular the Chacma baboon *Papio ursinus* (Ripamonti et al. 1997; Ripamonti et al. 2000; Ripamonti et al. 2008; Ripamonti and Roden 2010; Ripamonti *et al.* 2012). The induction of bone formation by hTGF-$\beta_1$, -$\beta_2$ and -$\beta_3$ in *Papio ursinus* opened a new avenue into the cascade of bone formation by induction whereby more focus is now on the understanding of how the signal transduction pathways inducing and modulating bone formation function so as to mechanistically understand the osteoinduction cascade.

Both BMP and TGF-β signaling pathways share a homology in how a signal migrates from the cell surface towards the nucleus (Miyazawa *et al.*, 2002). The only distinctive features that separate the two pathways into specific roles during cellular homeostasis are the Smad activated groups and promoter regions on the genome. Whilst BMPs mainly control cellular differentiation, specifically those of bone related cells, e.g. osteoblasts and osteoclasts, TGF-β is involved in many cellular processes especially those of cell growth and proliferation (Blobe *et al.,* 2000; Hanahan and Weinberg, 2000). TGF-β is a dimerized amino acid structure which in its final form produces a 25KDa active molecule with many conserved structural motifs in the main structure of all TGF-β isoforms (Herpin *et al.*, 2004). TGF-β has nine *cysteine* residues that are conserved among its family; eight form *disulfide bonds* within the molecule to create a *cysteine knot* structure characteristic of the TGF-β superfamily while the ninth cysteine forms a bond with the ninth cysteine of another TGF-β molecule to produce a dimer (Daopin *et al.*, 1992). Many other conserved residues in TGF-β are thought to form secondary structures through hydrophobic interactions. The region between the fifth and sixth conserved cysteines houses the most divergent area of TGF-β molecules and is the site in the peptide which is exposed. This small region of the protein is implicated as being the active site which associates with a specific TGF-β receptor and determines the specificity of the TGF-β isoform (Daopin *et al.*, 1992; Herpin *et al.*, 2004). The TGF-β isoform binds first with a type I receptor which in turn phosphorylates a receptor-regulated Smad (R-Smad). This complex molecule then further associates with a coSmad such as Smad4. These R-Smad/coSmad complexes migrate and finally accumulate in the nucleus where they act as transcription factors and modulate target gene expression (Miyazawa *et al.,* 2002).

Of the three mammalian TGF-β isoforms, the hTGF-$\beta_3$ has been shown to be the most active morphogen involved in bone regeneration when tested in the non-human primate *Papio ursinus* (Ripamonti *et al.* 2008; Ripamonti, 2010; Ripamonti *et al.* 2012). The highest amount of bone formation by the hTGF-$\beta_3$ isoform, to date, has been shown to occur in treated full thickness mandibular defects within 30 and 180 days after implantation of doses of 125 μg hTGF-$\beta_3$ per gram of allogeneic insoluble collagenous bone matrix as carrier; results showed complete *restitutio ad integrum* of the mandibular defects by day 30 and 90 after implantation (Ripamonti 2006a). It is still unclear how the hTGF-β isoforms initiate the cascade of bone differentiation by induction, as limited molecular evidence is available to properly explain these findings in primate models (Ripamonti and Roden 2010).

Moreover, the mammalian TGF-β isoforms are powerful inducers of endochondral bone in primates only (Ripamonti *et al.*, 2008; Ripamonti 2010; Ripamonti *et al.*, 2012); this animal and phylogenetic-dependent tissue induction and morphogenesis still needs to be assigned and it is largely unknown.

With the development of advanced quantitative techniques into gene transcription, especially quantitative Real Time Polymerase Chain (qRT-PCR) reaction and its enhancement through the MIQE guidelines (Bustin *et al.*, 2009), the mythos of TGF-$\beta_3$ and its involvement in the bone induction process is slowly being unraveled, particularly for calvarial bone regeneration. Preliminary data now available in our laboratories Bone Research Laboratory 2012, unpublished data) indicate that TGF-$\beta_3$ has a two way *modus operandi*. Not only does TGF-$\beta_3$ directly affect its own expression, by up regulating the transcription of its own exon region on the genome, but TGF-$\beta_3$ also appears to function indirectly by increasing native cell proliferation and/or differentiation (Bone Research Laboratory 2012, unpublished data). The TGF-$\beta_3$ isoform may cause hyper-deposition of bone by indirectly stimulating osteoblasts to undergo mass mitosis which enables more osteogenic material to be synthesized and secreted (Bone Research laboratory 2012, unpublished data).

These findings to date have been restricted to the calvarium of the non-human primate where hTGF-$\beta_3$ appears to function in conjunction with other regulators and/or inhibitory ligands believed to originate from the *dura mater* and/or the underlying arachnoids, thus affecting the overall bone formation performance (Bone Research Laboratory, unpublished data and Ripamonti *et al.*, 2008; Ripamonti *et al.*, 2009c). Of note, the calvarial bone develops *via* intramembranous ossification as opposed to the endochondral ossification of long bones; it still remains to be seen whether the *modus operandi* of hTGF-$\beta_3$ is identical in the craniofacial bones *vs*. the long bones of endochondral origin, or whether in each bony site the hTGF-$\beta_3$ acts through parallel but different pathways to induce the cascade of bone differentiation.

## Tissue Induction and Regeneration by Striated Muscle Stem Cells

Tissue engineering using pluripotent embryonic stem cells comes with the challenges of tissue compatibility and the ethical dilemma of the use of human embryos. Thus, the focus of regenerative medicine has turned to the potential use of postnatal stem cells, and in particular striated muscle-derived stem cells (MDSCs). MDSCs fulfil all the criteria of stem cells, namely, the ability for self-renewal, multilineage differential potential and stem cell marker expression (Usas *et al.*, 2011). Residing in the vicinity of the basal lamina of capillaries where satellite cells are normally found (Lee *et al.*, 2000), these cells which remain quiescent are brought to the site of injury by the vasculature, whereupon their activity is initiated.

The pivotal role of angiogenesis in tissue repair is well documented with the rate of endothelial cell replication increasing significantly in response to injury (Schwartz and Benditt, 1977). Kovacic and Boehm (2009) reviewed mesoangioblasts, an embryonic group of progenitor cells which arise from the roof and lateral walls of the dorsal aorta. These cells which give rise to endothelium, skeletal muscle, bone and cartilage, also differentiate into pericytes (Minasi *et al.*, 2002). Thus a common ontogeny exists between these cells and tissues and an intimate relationship of pericyte to endothelial cell is established early in development (Minasi *et al.*, 2002). The perivascular stem cell niches that ultimately reside within multiple human organs (Crisan *et al.*, 2008) play a critical role in angiogenesis, a process modulated by paracrine signals (Betsholtz *et al.*, 2005) which is ultimately a prerequisite for tissue induction and morphogenesis.

Importantly, Crisan *et al* (2008) showed that multi organ mesenchymal stem cells (MSCs) are perivascular pericytic stem cells. They demonstrated that all MSCs are pericytes based on the presence of CD146+, CD34-, CD45- and CD56- markers on both pericytes and mesenchymal stem cells. The converse, however, is not the case – all pericytes are not MSCs. Furthermore, the limited differentiation capability of adult mesenchymal stem cells and pericytes, as illustrated by their inability to form teratomas in heterotopic sites, shows that they cannot be classified with embryonic stem cells (Crisan *et al.*, 2008; Fong *et al.*, 2010).

An elegant study by Young *et al.* (2001) revealed by means of clonogenic analysis the capacity of rat cells derived from skeletal muscle tissue for cell renewal. Although cloning of these cells was undertaken *in vitro* they retained their phenotypic expression. Furthermore, following induction of differentiation of these cells an insulin-dexamethasone analysis showed that phenotypic markers of skeletal muscle, cartilage and bone were expressed. Whether these cells directly differentiate into either osteogenic or chondrogenic lineages or first undergo a dedifferentiation process is still widely debated. Medici *et al.* (2010) proposed the dedifferentiation process of the pericyte into a stem cell to be mediated via an activin-like kinase-2 (ALK2) receptor. Patients with fibrodysplasia ossificans progessiva, a pathologic condition in which endochondral bone forms heterotopically, are known to carry a heterozygous activating mutation in the gene encoding ALK-2. Whilst TGF-$\beta_2$ and BMP-4 were shown to activate ALK-2 and promote endothelial to mesenchymal cell transformation, BMP-7 inhibited this transformation (Medici *et al.,* 2010).

Due to the heterogeneous nature of satellite cells, Lee *et al.* (2000) isolated the mc13 line following purification of skeletal muscle-derived stem cells from the *mdx* mouse. *In vitro* and *in vivo* studies of this clone showed its potential for both myogenic and osteogenic differentiation. To induce these cells to differentiate into the osteogenic line the addition of bone morphogenetic/osteogenic proteins (BMPs/OPs) is required (Lee *et al.*, 2000; Lee *et al.*, 2001). To eliminate the potential risks associated with direct delivery of BMPs/OPs *via* vectors such as retroviruses and adenoviruses, Lee *et al.* (2001) utilised *ex vivo* gene therapy techniques in their study where mice muscle-derived cells were transduced with adenovirus encoding the recombinant human BMP-2 gene. When implanted with collagen and muscle cells engineered to express BMP-2, critical–sized bone defects in mice calvaria showed almost complete healing after four weeks. Defect healing without BMP-2 was less than 50%. Localisation of osteocalcin in the newly formed bone by immunohistochemistry verified that differentiation of cells into the osteogenic lineage had occurred *in vivo*. In keeping with the requirements for osteoinduction, the MDSCs have a dual role, which is, providing a delivery vehicle for the soluble signal and being the source of cells that differentiate into osteoblasts (Lee *et al.*, 2000).

Several studies undertaken in the non-human primate *Papio ursinus* show the critical role of skeletal muscle derived stem cells in bone induction and morphogenesis (Ripamonti *et al.*, 2008, Ripamonti *et al.*, 2009a, Ripamonti *et al.*, 2009b, Teare *et al.*, 2008). Morcellated fragments of autogenous *rectus abdominis* muscle partially restored the endochondral osteoinductivity of the TGF-$\beta_3$ isoform when implanted in calvarial defects of adult baboons (Figure 6) (Ripamonti *et al.*, 2008). These results were replicated in studies on periodontal regeneration where harvested autogenous *rectus abdominis* muscle was finely minced, added to 75μg of hTGF-$\beta_3$ in Matrigel® matrix and implanted in surgically created class II and III furcation defects of *P. ursinus* (Ripamonti *et al.*, 2009a, Ripamonti *et al.*, 2009b). The direct application of 75μg of hTGF-$\beta_3$ in Matrigel matrix together with the addition of minced fragments of autogenous *rectus abdominis* muscle resulted in greater alveolar bone formation and cementogenesis when compared with periodontal tissue regeneration induced by the implantation of hTGF-$\beta_3$ solo in the Matrigel matrix (Figure 7) (Ripamonti *et al.*, 2009a, Ripamonti *et al.*, 2009b, Teare *et al.*, 2008).

In conclusion, therefore, the value of using skeletal muscle as a source of stem cells lies in the abundance of tissue source available within an individual, the ease with which this can be harvested, obviating the need for donor tissue from another individual, the latter presenting the risk of immunorejection, and notably, the ability of a single cell source having the ability to regenerate bone, periodontal ligament fibers and cementum (Ripamonti *et al.*, 2009a; Ripamonti *et al.* 2009b).

## AUTOGENOUS STRIATED MUSCLE CELLS AND THE INDUCTION OF CALVARIAL AND PERIODONTAL TISSUE REGENERATION

Systematic studies in our laboratory have shown that in the non-human primate *Papio ursinus* the three mammalian TGF-β isoforms, the recombinant hTGF-$\beta_1$, -$\beta_2$ and -$\beta_3$ proteins, induce rapid and substantial bone induction in heterotopic sites of the *rectus abdominis* muscle (Figures 4, 5) (Ripamonti *et al.*, 1997; Ripamonti *et al.,* 2000, Ripamonti *et al.* 2008; Ripamonti and Roden 2010). However, equal or higher doses of the recombinant proteins implanted orthotopically in calvarial defects do not induce bone formation (Figure 6) (Ripamonti *et al.* 1996; Ripamonti *et al.*, 2000; Ripamonti *et al.* 2008). On day 90, implantation of the mammalian TGF-β isoforms results in limited bone formation, pericranially only just below the *temporalis* muscle, with minimal bone formation at the edges of the craniotomies (Figure 6) (Ripamonti *et al.*, 2000; Ripamonti *et al.*, 2008).

Reverse transcription-polymerase chain reaction (PCR), Western and Northern blot analyses of tissue specimens generated by the hTGF-$\beta_3$ osteogenic device demonstrated robust expression of *Smads-6* and *-7* in orthotopic calvarial sites on day 30 (Ripamonti *et al.*, 2008). On day 90, the elevated expression of *Smad-6* and *-7* in orthotopic calvarial samples was not observed (Ripamonti *et al.*, 2008). The relative reduction of expression of both *Smad-6* and *-7* in calvarial sites as shown on day 90 correlated with the induction of bone formation pericranially on day 90 (Ripamonti *et al.,* 2008). We have previously suggested that *Smad-6* and *-7* expression in treated-calvarial defects may be due to the vascular endothelial network of the arachnoids and leptomeninges below expressing signaling proteins modulating the expression of the inhibitory Smads in pre-osteoblastic and osteoblastic cell lines regulating the induction of bone formation in the primate calvarium (Ripamonti *et al.*, 2008; Ripamonti *et al.*, 2009c).

The repetitive observation of predictable induction of large ossicles by the hTGF-$\beta_3$ in heterotopic sites of the *rectus abdominis* muscle suggested that the *rectus abdominis* muscle is endowed with multiple osteogenic stem cells niches capable of receptor activation and phosphorylation when ligated by the hTGF-β isoforms, in particular the hTGF-$\beta_3$ protein. Indeed, the addition of morcellated fragments of autogenous *rectus abdominis* muscle resulted, on one hand, in greater bone formation and deposition within the implanted calvarial defects but, on the other hand, in greater expression of the *Smad-6* and *-7* (Ripamonti, 2008). Importantly, however, the addition of morcellated fragments of autogenous *rectus abdominis* muscle partially restored the induction of bone formation mostly shown on day 90 with also the induction of bone on previously unreported endocranial sites (Figure 6) (Ripamonti, 2008; Ripamonti *et al.*, 2009c).

Tissue specimens harvested on day 90 showed often the presence of newly induced mineralized bone pericranially below the *temporalis* muscle, possibly an indication that muscle tissue may be a source of responding stem cells which would support and possibly restore the bone induction cascade as initiated by the recombinant hTGF-β isoforms. Additional calvarial studies were therefore designed to incorporate morcellated muscle tissue harvested from autogenous *rectus abdominis* muscle combined with the hTGF-$\beta_3$ osteogenic device (Ripamonti *et al.*, 2008). The induction of substantial bone formation, greater than hTGF-$\beta_3$/osteogenic devices solo (Figure 6), indicated that the harvested and morcellated striated muscle retains responding stem cells promptly capable of differentiation into osteoblastic-like cells (Ripamonti *et al.*, 2008; Ripamonti *et al.*, 2009c). Of interest, calvarial sections prepared on day 90 after implantation of hTGF-$\beta_3$-osteogenic devices with morcellated fragments of autogenous *rectus abdominis* muscle also showed the induction of chondrogenesis (Ripamonti *et al.*, 2008). Importantly thus, the addition of morcellated fragments of autogenous *rectus abdominis* muscle engineers endochondral bone formation with large islands of chondrogenesis as a recapitulation of embryonic development, even if implanted in calvarial defects, where the induction of bone formation is only membranous, without a chondrogenic phase (Ripamonti *et al.* 2008).

The complex tissue morphologies of the periodontal tissues, the locking of the teeth into the alveolar bone proper, the induction of cementogenesis along the root surfaces with embedded *bona fide* Sharpey's fibers are a superb example of Nature's molecular and morphological design and architecture. Indeed, the challenging theme of the complex tissue morphologies of the periodontal tissues is the molecular basis of morphogenesis and the induction of cementogenesis, which provide the cemental avascular layered tissue for the insertion of periodontal ligament fibers thus locking the tooth into the alveolar bone allowing masticatory forces not otherwise possible.

Periodontal tissue regeneration is the final goal of periodontal therapy. The three major challenges of periodontal tissue engineering as identified almost two decades ago (Ripamonti and Reddi, 1994) still need to be resolved so as to engineer periodontal tissue regeneration with the induction of newly formed cementum and the genesis of *bona fide* Sharpey's fibers, the essential ingredient to engineer periodontal tissue regeneration (Ripamonti and Reddi, 1994; Ripamonti, 2007).

The osteogenic proteins of the TGF-β supergene family induce *de novo* endochondral bone formation as a recapitulation of embryonic development and act as soluble signals for tissue morphogenesis sculpting the multicellular mineralized structures of the periodontal tissues with functionally oriented periodontal ligament fibers inserting into newly formed cementum (Ripamonti, 2007).

The observation that morcellated fragments of autogenous *rectus abdominis* muscle partially restored the inductive activity of hTGF-$\beta_3$ osteogenic devices when implanted in calvarial defects of the non-human primate *Papio ursinus* (Figure 6) (Ripamonti *et al.*, 2008; Ripamonti *et al.*, 2009c) has been replicated in periodontal regenerative studies using doses of the hTGF-$\beta_3$ osteogenic device combined with morcellated fragments of autogenous *rectus abdominis* muscle implanted in Class II and III furcation defects of *Papio ursinus* (Teare *et al.*, 2008; Ripamonti *et al.*, 2009a; Ripamonti and Petit, 2009). The direct application of 75μg hTGF-$\beta_3$ in Matrigel® matrix with the addition of morcellated fragments of autogenous *rectus abdominis* muscle resulted in greater alveolar bone formation (Teare *et al.*, 2008). Further studies showed that the addition of morcellated fragments of autogenous *rectus abdominis* muscle resulted in superior cementogenesis along the exposed root surfaces (Figure 7) (Ripamonti *et al.*, 2009a) when compared with the induction of periodontal tissue regeneration generated by the hTGF-$\beta_3$ osteogenic device solo in Matrigel® matrix (Teare *et al.*, 2008; Ripamonti *et al.*, 2009a; Ripamonti and Petit, 2009).

By simply morcellating fragments of autogenous *rectus abdominis* biopsies, we have shown that the striated *rectus abdominis* muscle is an important source of myoblastic stem cells that can be rapidly prepared and transplanted in non-healing calvarial and periodontal defects of the non-human primate *Papio ursinus* (Ripamonti *et al.*, 2009c).

Importantly, the above results using seemingly crude preparations of myoblastic/pericytic stem cells contained in morcellated fragments of *rectus abdominis* striated muscle, have indicated that the striated muscle retains responding mesenchymal stem cells capable of transformation into desired cellular phenotypes, that is, osteoblastic and cementoblastic cell lines, respectively, when in contact with specific extracellular matrix substrata, thus engineering calvarial (Ripamonti *et al.*, 2008; Ripamonti *et al.*, 2009c) and periodontal tissue induction and regeneration (Ripamonti *et al.*, 2009a; Ripamonti and Petit, 2009).

## Periodontal Tissue Induction by Matrigel® Matrix, Transforming Growth Factor-B3, and Morcellated *Rectus Abdominis* Muscle in Furcation Defects of *Macaca Mulatta*

The regenerative potential of bone, a highly vascular three-dimensional mineralized matrix which remodels throughout life and heals without scarring has been known since antiquity (Reddi 1981; Reddi 1984). Despite the regenerative capacity of bone, periodontal osseous defects, however, lack the template for orchestrated tissue regeneration. Periodontal tissue engineering has often proved to be elusive because of the challenges involved in differentiation, migration, attachment and spatial and temporal positioning of a variety of embryologically different cell types on a supportive avascular and mineralized dentinal substratum (Ripamonti and Reddi 1994; Ripamonti 2007). Cellular activity is a vital component of comparative large bone defects' repair and regenerative processes. Striated muscle containing several different stem cell 'niches' harboring a variety of stem cells with osteogenic and cementogenic phenotypic differentiation pathways could be an adjuvant to soluble molecular signals to induce regeneration in Class II and III furcation defects as a result of recurrent episodes of chronic advanced periodontitis.

Since the early nineties, our laboratory's approach has focused on regenerative phenomena invocated by the osteogenic proteins of the TGF-β supergene family in Class II and III furcation defects of the adult non-human primate *Papio ursinus* (Ripamonti *et al.*, 1994; Ripamonti and Reddi, 1994; Ripamonti *et al.*, 1996; Ripamonti and Reddi 1997; Ripamonti *et al.*, 2002; Ripamonti, 2007). Results after short and long term studies have shown unequivocally that naturally-derived highly purified BMPs/OPs, recombinant hOP-1 and hTGF-$\beta_3$ when implanted in Class II and III furcation defects of *Papio ursinus* induce cementogenesis with the insertion of functionally oriented periodontal ligament fibers cursing within a newly formed highly vascular periodontal ligament system with Sharpey's fibers *de novo* generated within the newly secreted as yet to be mineralized cementoid matrix (Ripamonti 2007; Ripamonti *et al.*, 2009b).

In more recent experiments in adult *Macaca mulatta* monkeys (Bone Research Laboratory 2012 unpublished data), our laboratories have used Class II furcation defects as tissue engineering bioreactors (Stevens *et al.*, 2005) to construct the induction of periodontal tissue regeneration after challenging the surgically prepared bioreactors with 75μg hTGF-$\beta_3$ recombined with Matrigel® matrix with or without the addition of morcellated fragments of autogenous *rectus abdominis* striated muscle, as previously reported in furcation defects of *Papio ursinus* (Ripamonti *et al.*, 2009a). To further study the cascade of tissue induction and morphogenesis, regenerated periodontal tissues, the cementum and the alveolar bone, were harvested at two time periods for molecular analyses: on the day of implantation (day 0), and again 60 days after implantation (day 60), just before euthanasia and tissues harvest for undecalcified histology. RNA extracted from the cementum and alveolar bone was subjected to reverse transcription real time polymerase chain reaction (RT-PCR) to determine the relative-fold difference in the target gene of samples in tissues harvested on day 0 and 60. Cementogenic and osteogenic markers including osteocalcin (OC), cementum protein 1 (CEMP 1), OP-1, BMP-2 and TGF-$\beta_3$ in both cementoid and mineralized alveolar bone matrices were analyzed. The results showed that OC, CEMP 1, OP-1, BMP-2 and TGF-$\beta_3$ mRNAs were all expressed to a varying degree in cementoid extracts at both time periods (Bone Research Laboratory 2012 unpublished data).

OC gene product expression raised several fold after the first periodontal instrumentation with continuous expression during periodontal tissue regeneration. Of note, OP-1 gene product is expressed within the cementoid matrix regulating *in vivo* cementogenesis during periodontal tissue regeneration (Bone Research Laboratory 2012 unpublished data). OP-1 is a critical regulatory gene involved in self-repair and self-inductive phenomena of both cementoblasts and periodontal ligament cells to induce and maintain cementogenesis with functionally oriented periodontal ligament fibers. Of note, OP-1 in cementum showed a two-fold increase as compared to BMP-2 at both time periods (Bone Research Laboratory 2012 unpublished data). The molecular data reflect the morphological observation of tissue induction as reported after implantation of hOP-1 in periodontal osseous defects of both canine and non-human primates models (Ripamonti, 2007); hOP-1 preferentially induces cementogenesis in the context of periodontal tissue regeneration (Ripamonti 2007); to the contrary, hBMP-2 when implanted in periodontal furcation defects is preferentially osteogenic with limited if any induction of cementogenesis (Ripamonti 2007; Ripamonti and Petit 2009; Ripamonti *et al.*, 2009b). Importantly, RT-PCR of regenerated alveolar bone samples showed that OP-1 and BMP-2 gene products were expressed at both time periods in specimens of Matrigel® matrix combined with morcellated *rectus abdominis* muscle, irrespective of the addition of the TGF-$\beta_3$ isoform (Bone Research Laboratory 2012 unpublished data).

After 60 days of healing, Class II mandibular furcation defects treated with Matrigel® matrix with or without 75μg hTGF-$\beta_3$, and with and without morcellated fragments of *rectus abdominis* muscle, showed cementogenesis and alveolar bone regeneration to a varying degree according to the treatment modalities (Figure 8).

Significant periodontal tissue regeneration was shown by furcation defects treated with Matrigel® matrix with morcellated fragments of *rectus abdominis* muscle with or without hTGF-$\beta_3$, highlighting the role of the extracellular matrix and responding myoblastic/myoendothelial/pericytic stem cells during regenerative periodontal surgical procedures (Figure 8).

Furcation defects treated with Matrigel® matrix with morcellated fragments of *rectus abdominis* muscle and furcation defects treated with Matrigel® matrix, morcellated fragments of *rectus abdominis* muscle plus the addition of 75μg hTGF-$\beta_3$ showed comparatively similar results (Figure 9), with the exclusion of osteoid matrix, which was greater in specimens of Matrigel® matrix without hTGF-$\beta_3$. Of note, furcation defects treated with morcellated fragments of *rectus abdominis* muscle with or without the addition of 75μg hTGF-$\beta_3$ showed pronounced alveolar bone regeneration and cementogenesis along the surgically denuded root surfaces (Figures 8, 9).

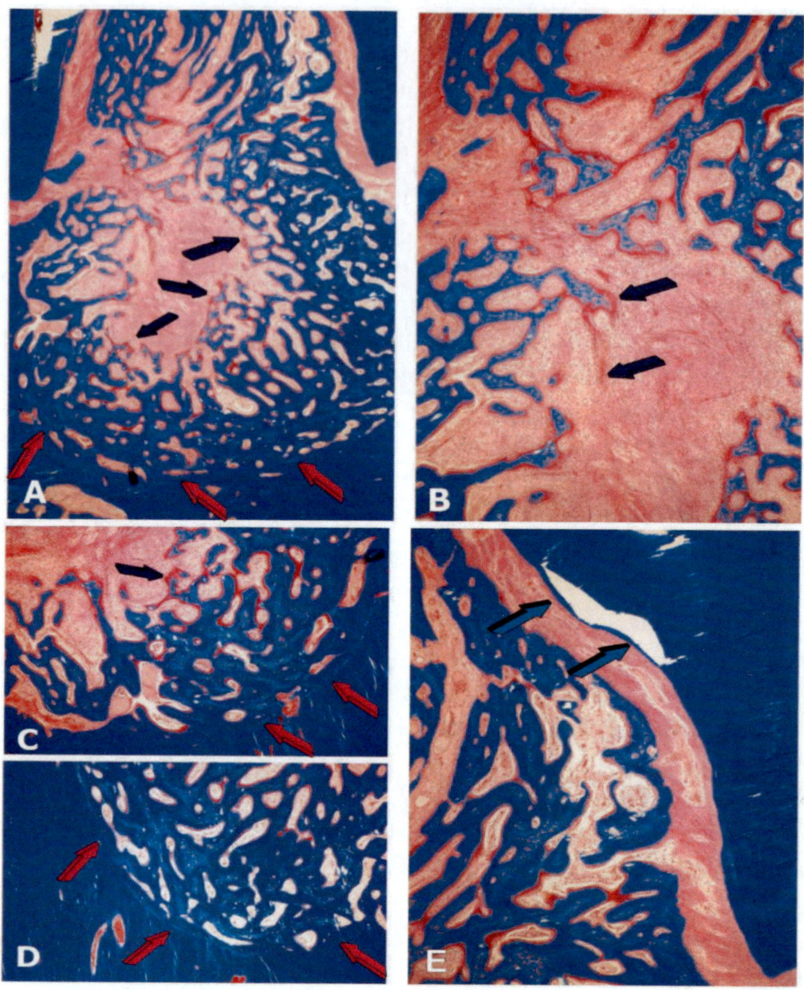


Figure 8. Induction of mineralized newly formed bone in ***blue*** surfaced by continuous osteoid seams populated by contiguous osteoblasts in mandibular Class II furcation defects of *Macaca mulatta* monkeys implanted with morcellated fragments of autogenous *rectus abdominis* muscle blended in Matrigel® matrix. (A,C,D): Prominent alveolar bone regeneration extending coronally from the apically prepared furcation defects at the residual bony housing (*magenta* arrows); (A,B): Prominent osteogenesis with osteoid seams (dark ***blue*** arrows) surfacing newly formed trabeculae of mineralized bone surrounding highly vascular and cellular mesenchymal tissue interpreted as the tissue response to the implantation of finely minced fragments of *rectus abdominis* muscle, rapidly transforming into secreting osteoblastic-like cells. (E): Detail of coronally located instrumented root surface with newly formed cementum (light ***blue*** arrows). Undecalcified sections cut at 7μm stained free floating with Goldner's trichrome.

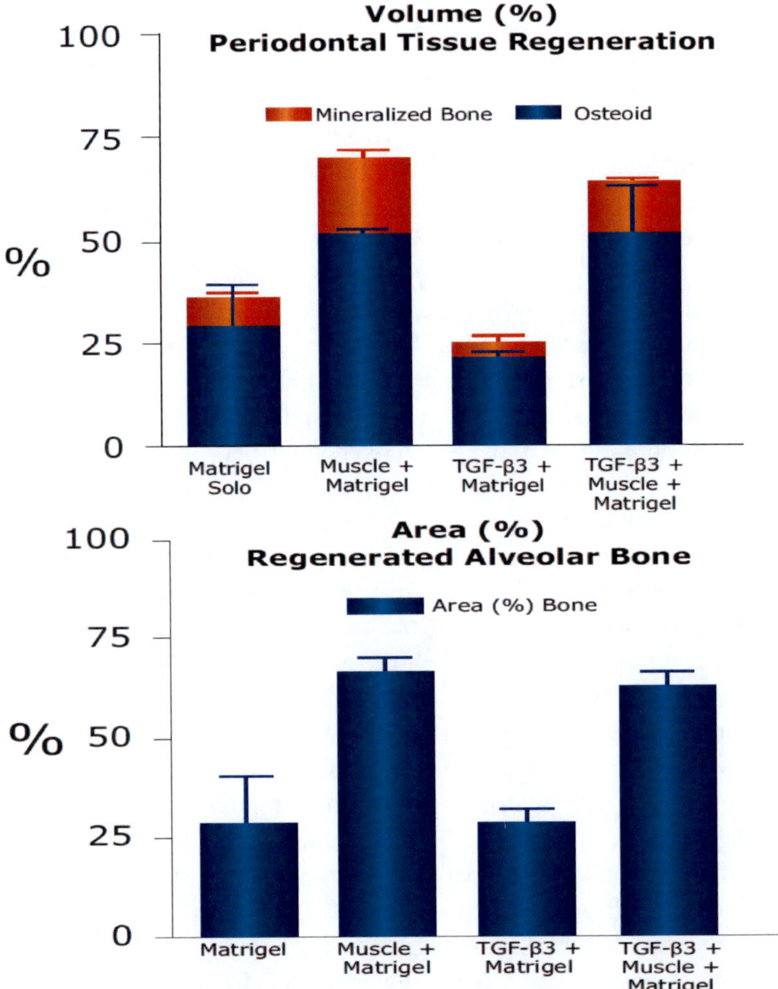


Figure 9. Periodontal tissue regeneration in Class II furcation defects surgically prepared in *Macaca mulatta* monkeys after implantation of Matrigel® matrix combined with morcellated fragments of autogenous *rectus abdominis* muscle with and without human recombinant transforming growth factor-$\beta_3$ (hTGF-$\beta_3$); of note, the substantial induction of osteoid (volume %) by defects treated with Matrigel® matrix plus muscle cells but without hTGF-$\beta_3$; Matrigel® matrix with morcellated fragments of autogenous *rectus abdominis* muscle with or without hTGF-$\beta_3$ induced greater alveolar bone regeneration (volume %) and regenerated alveolar bone area (area %) when compared to Matrigel® matrix with or without hTGF-$\beta_3$.

Histological analyses showed the apical extent of the rotary instrumentation well beyond the apical position of the root surfaces (Figure 8). Furcation defects treated with morcellated fragments of *rectus abdominis* muscle showed a reproducible pattern of tissue induction and morphogenesis as illustrated in Figure 8. This was characterized by highly cellular and vascularized mesenchymal tissue surrounded by prominent angiogenesis with multiple mineralized trabeculae of newly formed woven bone covered by large osteoid seams (Figure 8B). Highly vascularized, highly cellular poorly organized mesenchymal tissues within newly formed bone were only generated in furcation defects treated with morcellated fragments of *rectus abdominis* muscle with or without the addition of 75μg hTGF-$\beta_3$ (Figure 8).

Striated muscle is known to harbour myoblastic/myoendothelial and perivascular/ pericytic stem cells capable of rapid differentiation into secreting osteoblast-like bone forming cells (Kovacic and Boehm, 2009; Zheng *et al.*, 2007). The generation of woven osteogenetic fronts within the implanted furcation defects with Matrigel® matrix and morcellated *rectus abdominis* fragments with or without the hTGF-β3 isoform, possibly represents the rapid differentiation and transformation of myoblastic/myoendothelial and perivascular/ pericytic stem cells into osteoblastic bone forming cells from the implanted morcellated fragments of the implanted *rectus abdominis* muscle. The morcellated fragments of autogenous *rectus abdominis* muscle might have converged to the centre of the treated furcation defects primarily providing a highly differentiating stem cell 'niche' within the micro-environment of the implanted furcation defects acting as bone forming bioreactors after implantation of Matrigel® matrix with and without the hTGF-$\beta_3$ isoform.

Although stem cells were delivered by fundamentally crude morcellated *rectus abdominis* muscle preparations, comparative histomorphometrical analyses of morcellated fragments co-implanted with hTGF-$\beta_3$ osteogenic devices in a variety of different biological micro-environments induced greater cementogenesis and alveolar bone regeneration (Ripamonti *et al.,* 2009a; Ripamonti *et al.,* 2009c) not least the induction of bone formation in non-healing calvarial defects of *Papio ursinus* (Ripamonti *et al.*, 2008).

The functionality of engineered periodontal tissues has been evaluated primarily by the morphological and morphometrical evidence of cementogenesis with functionally oriented periodontal ligament fibers (Bartold *et al.*, 2000; Ripamonti *et al.*, 2009b). We have already proposed that the induction of bone formation, together with the induction of cementogenesis, should be based on the expression patterns of selected gene products as a recapitulation of embryonic development (Ripamonti, 2007; Ripamonti *et al.*,

2009b). In spite of the abundance of studies at morphological level, to date, the molecular and cellular mechanisms that set the *in vivo* postnatal induction of periodontal tissue regeneration and the initiation of cementogenesis are still unknown. Our preliminary data obtained in *Macaca mulatta* monkeys showed comparative morphological and molecular data and show that OC and OP-1 gene products modulate cementogenesis in *Macaca mulatta* periodontal bioreactors challenged with and without hTGF-$\beta_3$ in Matrigel® matrix with morcellated *rectus abdominis* muscle fragments. OC gene product expression was raised several fold after the first periodontal instrumentation, with continuous expression during periodontal tissue regeneration. The molecular data showed that OP-1 gene product is expressed within the cementoid matrix regulating *in vivo* cementogenesis during periodontal tissue regeneration. OP-1 is a critical regulatory gene involved in self-repair and self-inductive phenomena of both cementoblasts and periodontal ligament cells to induce and maintain cementogenesis with functionally oriented fibers (Amar *et al.*, 1997; Hakki *et al.*, 2010; Ripamonti, 2007; Ripamonti and Petit, 2009; Ripamonti *et al.*, 2009b).

The above preliminary molecular results in *Macaca mulatta* obtained in a different animal model than the multi-tested Chacma baboon *Papio ursinus*, have shown that Matrigel® matrix further activated by autogenous morcellated *rectus abdominis* striated muscle, initiates cementogenesis with inserted newly generated periodontal ligament fibers with or without the addition of doses of the recombinant hTGF-$\beta_3$ isoform. Importantly, Matrigel® matrix loaded with morcellated fragments of striated muscle could become a routine therapeutic approach after fragmentation of striated muscle harvested within the oral cavity. This may help design therapeutic strategies based on cell biology of matrix-cell interactions to induce reproducible cementogenesis in clinical contexts.

## Challenges and Perspectives of Muscle Cell Transplantation and Regeneration in Bone Tissue Engineering

Current advances in the realm of molecular, cellular and developmental biology, tissue biology and experimental reconstructive surgery have rapidly reached a previously unknown biological knowledge of the mechanistic molecular mechanisms of tissue induction and morphogenesis (Reddi, 2000;

Ripamonti, 2006; Ripamonti, 2010). This explosive knowledge has cut the boundaries between several different disciplines linking this novel biological and surgical knowledge into the emerging science of tissue engineering and regenerative medicine (Reddi, 1994; Reddi, 2000; Ripamonti, 2006; Ripamonti, 2010).

Importantly, the mechanistic understanding of the fascinating phenomenon of "*Bone: Formation by Autoinduction*" (Urist, 1965; Reddi and Huggins 1972) has set the rules of the tissue engineering paradigm as the induction of bone formation by combinatorial molecular protocols to restore and reconstitute the biological activity of the osteogenic soluble molecular signals of the TGF-β supergene family (Reddi 2000; Ripamonti *et al.,* 2004; Ripamonti, 2006; Ripamonti, 2010). Insoluble signals or substrata only when recombined with soluble molecular signals trigger the ripple-like cascade of tissue induction and morphogenesis (Reddi, 2000; Ripamonti *et al.*, 2004); the morphogenetic insoluble extracellular signals are critically regulated both in time and space and finely tuned by a vast network of inhibitors and activators (Reddi 2000; Groppe 2008; Ripamonti *et al.,* 2009c).

Tissue engineers and skeletal reconstructionists alike realized however that the induction of bone formation in clinical contexts is on a different scale altogether when compared to pre-clinical data that also include non-human primate species; indeed even non-human primate species may not adequately translate and reproduce morphogens-related therapeutic responses in *Homo sapiens* (Ripamonti 2010; Ripamonti *et al.,* 2012). During the last International Conference on Bone Morphogenetic Proteins (Lake Tahoe June 19-23 2012), a session was eventually set to discuss pre-clinical results obtained to date in canine and non-human primates models *vs*. the uninspiring results obtained thus far in clinical contexts (Ripamonti 2012. Driving the induction of bone formation in pre-clinical contexts by the soluble osteogenic molecular signals of the TGF-β supergene family; Ferretti and Ripamonti 2012. A critical appraisal of human bone tissue engineering).

Our laboratories are mostly supporting the evolutionary loss of tissue regeneration in different animal species (Bely and Nyberg 2010; Bely 2010) that support important evolutionary differences in regenerative potential between species rather than focusing on biotechnologically tailoring recombinant osteogenic devices altering pro- and pre-domain sequences to yield recombinant proteins escaping inhibitory processes by a number of protein modulators including the noggin proteins (Ripamonti *et al.,* 2009c).

The conceptual formulation of the structure/activity profile as the biological basis of apparent redundancy of the several osteogenic molecular isoforms of the BMPs/OPs and TGF-β family's needs now to be assigned, and molecularly and morphologically dissected to control selected tissue morphogenesis in pre-clinical and clinical contexts. Periodontal tissue morphogenesis and the induction of cementogenesis along exposed root surfaces has dramatically shown the structure/activity profile of molecularly homologous but biologically different isoforms on the induction of periodontal tissue regeneration (Ripamonti 2006; Ripamonti 2007). hBMP-2 applied singly to periodontal defects of canine and primate models is preferentially osteogenic but not cementogenic (Ripamonti, 2007; Ripamonti *et al.,* 2009); on the contrary, hOP-1, when in contact with dentine extracellular matrices, is preferentially cementogenic when evaluated at days 60 and 180 after implantation in exposed furcation defects of the non-human primate *Papio ursinus* (Ripamonti *et al.*, 1996; Ripamonti 2007; Ripamonti *et al.,* 2009).

In preclinical studies in a canine model, root surfaces treated with doses of hBMP-2 showed that cementum regeneration was less than control treatment without hBMP-2 (Choi *et al.,* 2002). Recombinant hBMP-2 does not have a significant effect on cementum regeneration and formation of a functionally oriented periodontal ligament system (Sigurdsson *et al.*, 1995; Giannobile *et al.*, 1998; Choi *et al.,* 2002). Mechanistically, the structure/activity profile of recombinant hBMP2 inhibits differentiation and mineralization of cementoblasts (Zhao *et al.,* 2003). Cementoblasts exposed *in vitro* to hBMP-2 significantly reduce bone sialoprotein and collagen type I gene expression with inhibition of cell-induced mineral nodule formation (Zhao *et al.,* 2003).

The addition of myoblastic, pericytic and myoendothelial stem cells contained in multiple niches of the striated muscle has shown in pre-clinical studies superior osteogenesis and cementogenesis in calvarial and periodontal osseous defects of *Papio ursinus* (Ripamonti *et al.,* 2008; Ripamonti *et al.,* 2009c; Ripamonti *et al,.* 2009a).

Ultimately, translational medicine in clinical contexts has so far suggested, if not now openly requested, to reconsider and re-evaluate the tissue engineering results obtained so far in clinical contexts. Research scientists across multiple disciplines do not know – or – do not grasp as yet – if regenerative medicine, that has often descended at times, "*prove to be a nadir out of which only success can emerge, or is it a subject so fatally flawed by a misappropriation of medical principles and commercial hype that it can only serve to deceive and ultimately fail*" (Williams, 2006). Though the principles of tissue engineering and regenerative medicine at large have resulted in the appropriation of a widely novel and hyperextended biological knowledge of tissue and molecular biology, the fact remains that none of the tissue engineering strategies so magnificently implemented and mechanistically resolved at molecular level in pre-clinical contexts are routinely translated in clinical contexts (Williams, 2006; Ripamonti *et al.*, 2012).

The critical challenge of regenerative medicine is to start to identify systematically the molecular and cellular basis responsible for significant differences in regenerative potential amongst animal species and animal phyla (Ripamonti, 2010). As challengingly stated by Alexandra Bely (2010), "*the evolutionary loss of regenerative ability represents a fundamental and perplexing problem in biology*". The extensive variation in regenerative capabilities across animal phyla is also not easily explained (Bely and Nyberg 2010). As previously suggested (Ripamonti, 2010) major research efforts should now be devoted to analyze genetically the mammalian wound healing trait controlling the extent of tissue induction and regeneration (McBrearty *et al.*, 1998; Ripamonti, 2010).

## ACKNOWLEDGMENTS

Our studies on the "bone induction principle" have been constantly supported by the University of the Witwatersrand, Johannesburg, the Faculty of Health Sciences Medical School, the Medical Research Council of South Africa, and the National Research Foundation since the late eighties when undecalcified sections cut by Barbara van den Heever have shown the induction of bone formation in the Chacma baboon *Papio ursinus*, as well as the induction of periodontal tissue regeneration and cementogenesis on uniquely cut undecalcified sections at 3 μm of the root dentine/periodontal interface. The histological and molecular experimentation in *Macaca mulatta* has been supported by a specific grant of the red fund of the South African Dental Association.

## REFERENCES

Amar, S; Chung, KM; Nam, SH; Karatzas, S; Myokai, F; Van Dyke, TE. Markers of bone and cementum formation accumulate in tissues regenerated in periodontal defects treated with expanded polytetrafluoroethylene membranes. J Periodont Res, 1997, 32, 148-158.

Bartold, PM; McCulloch, CA; Narayanan, AS; Pitaru, S. Tissue engineering: a new paradigm for periodontal regeneration based on molecular and cell biology. *Periodontol 2000*, 2000, 24, 253-269.

Bely, AE; Nyberg, KG. Evolution of animal regeneration: re-emergence of a field. *Trends in Ecology and Evolution*, 2010, 25, 161-170.

Bely, AE. Evolutionary loss of animal regeneration: pattern and process. Integr Comp Biol, 2010, 50, 515-527.

Betsholtz, C; Lindblom, P; Gerhardt, H. *Role of pericytes in vascular morphogenesis. Exs 2005*, 94, 115-125.

Blobe, GC; Schiemann, WP; Lodish, HF. Role of transforming growth factor beta in human disease. *N. Eng. J. Med.,* 2000, 342, 1350–1358.

Bridges, JB; Pritchard, JJ. Bone and cartilage induction in the rabbit. *J. Anat.,* 1958, 92, 28-38.

Bustin, SA; Benes, V; Garson, JA; Hellemans, J; Huggett, J; Kubista, M; Mueller, R; Nolan, T; Pfaffl, MW; Shipley, GL; Vandesompele, J; Wittwer, CT. The MIQE Guidelines: Minimum information for publication of Quantitative Real-Time PCR experiments. *Clinical Chem.*, 2009, 55, 611-622.

Caplan, AI. All MSCs are Pericytes? *Cell Stem Cell*, 2008, 3, 229-230.

Chen, CW; Montelatici, E; Crisan, M; Coselli, M; Huard, J; Lazzari, L; Peault, B. Perivascular multi-lineage progenitor cells in human organs: regenerative units, cytokine sources or both? *Cytokine. Growth Factor Rev.,* 2009, 20, 429-434.

Choi, S-H; Kim, C-K; Cho, K-S; Huh, JS; Sorenson, RG; Wozney, JM; Wikesjö, UM. Effect of recombinant human bone morphogenetic protein-2/absorbable collagen sponge (rhBMP-2/ACS) on healing in 3-wall intrabony defects in dogs. *J. Periodontol.*, 2002, 73, 63-72.

Corsi, KA; Schwarz, EM; Mooney, DJ; Huard J. Regenerative medicine in orthopaedic surgery. *J. Orthop. Res.,* 2007, 25, 1261-1268.

Crisan, M; Yap, S; Casteilla, L; Chen, CW; Corselli, M; Park, T S; Andriolo, G; Sun, B; Zheng, B; Zhang, L; Norotte, C; Teng, PN; Traas, J; Schugar, R; Deasy, BM; Badylak, S; Buhring, HJ; Giacobino, JP; Lazzari, L; Huard, J; Peault, B. A perivascular origin for mesenchymal stem cells in multiple human organs. *Cell Stem Cell*, 2008, 3, 301-313.

Daopin, S; Piez, K; Ogawa, Y; Davies, D. Crystal structure of transforming growth factor-beta 2: an unusual fold for the superfamily. *Science,* 1992, 257, 369–373.

Duneas, N; Crooks, J; Ripamonti, U. Transforming growth factor-β1: induction of bone morphogenetic protein genes expression during endochondral bone formation in the baboon, and synergistic interaction with osteogenic protein-1 (BMP-7). *Growth Factors*, 1998, 15, 259-277.

Ferretti, C; Ripamonti, U. Human segmental mandibular defects treated with naturally derived bone morphogenetic proteins. *J. Craniofac. Surg.*, 2002, 13, 434-444.

Ferretti, C; Ripamonti, U. A critical appraisal of human bone tissue engineering. International conference on bone morphogenetic proteins June19-23 2012 Lake Tahoe, California.

Fong CY, Gauthaman K, Bongso A. Teratomas from pluripotent stem cells: a clinical hurdle. *J. Cell Biochem.,* 2010, 111, 769-781.

Friendlander, GE; Perry, CR; Cole, JD; Cook, SD; Cierny, G; Muschler, GE; Zych, GA; Calhoun, JH; LaForte, AJ; Yin, S. Osteogenic protein-1 (bone morphogenetic protein-7) in the treatment of tibial nonunions. *J. Bone Joint Surg.*, 2001, 83, 151-S158.

Garrison, KR; Donell, S; Ryder, J; Shemilt, I; Mugford, M; Harvey, I; Song, F. Clinical effectiveness and cost-effectiveness of bone morphogenetic proteins in the non-healing of fractures and spinal fusion: A systematic review. *Health Technol. Assess.*, 2007, 11, 1-150, iii-iv.

Gautschi, OP; Frey, SP; Zellweger, R. Bone morphogenetic proteins in clinical applications. *ANZ J. Surg.,* 2007, 77, 626-631.

Giannobile, WV; Ryan, S; Shih, MS; Su, DL; Kaplan, PL; Chan, TC. Recombinant human osteogenic protein-1 (OP-1) stimulates periodontal wound healing in class III furcation defects. *J. Periodontol.*, 69, 129-137.

Govender, S; Csimma, C; Genant, HK; Valentin-Opran, A; Amit, Y; Arbel, R; Aro, H; Atar, D; Bishay, M; Borner, MG; Chiron, P; Choong, P; Cinats, J; Courtenay, B; Feibel, R; Geulette, B; Gravel, C; Haas, N; Raschke, M; Hammacher, E; Van Der Velde, D; Hardy, P; Holt, M; Josten, C; Ketterl, RL; Lindeque, B; Lob, G; Mathevon, H; Mccoy, G; Marsh, D; Miller, R; Munting, E; Oevre, S; Nordsletten, L; Patel, A; Pohl, A; Rennie, W; Reynders, P; Rommens, PM; Rondia, J; Rossouw, WC; Daneel, PJ; Ruff, S; Ruter, A; Santavirta, S; Schildhauer, TA; Gekle, C; Schnettler, R; Segal, D; Seiler, H; Snowdowne, RB; Stapert, J; Taglang, G; Verdonk, R; Vogels, L; Weckbach, A; Wentzensen, A; Wisniewski, T. Recombinant human bone morphogenetic protein-2 for treatment of open tibial fractures: A prospective, controlled, randomized study of four hundred and fifty patients. *J. Bone Joint Surg. Am.,* 2002, 84-A, 2123-2134.

Groppe, J; Hinck CS, Samavarchi-Tehrani, P; Zubieta, C; Schuermann, JP; Taylor, AB; Schwarz, PM; Wrana, JL; Hinck, AP. Cooperative assembly of TGF-beta superfamily signaling complexes is mediated by two disparate mechanisms and distinct modes of receptor binding. *Mol. Cell,* 2008, 29:157-168.

Hakki, SS; Foster, BL; Nagatomo, KJ; Foster, BL; Nagatomo, KJ; Bozkurt, SB; Hakki, EE; Somerman, MJ; Nohutcu, RM. Bone morphogenetic protein-7 enhances cementoblast function in vitro. *J. Periodontol.*, 2010, 81, 1663-1674.

Hanahan, D; Weinberg, RA. The hallmarks of cancer. Cell, 2000 , 100, 57–70.

Herpin, A; Lelong, C; Favrel, P.Transforming growth factor-beta-related proteins: an ancestral and widespread superfamily of cytokines in metazoans. *Dev. Comp. Immunol.,* 2004, 28, 461–485.

Kovacic, JC; Boehm, M. Resident vascular progenitor cells: An emerging role for non-terminally differentiated vessel-resident cells in vascular biology. *Stem Cell Res*., 2009, 2, 2-15.

Lee, JY; Qu-Petersen, Z; Cao, B; Kimura, S; Jankowski, R; Cummins, J; Usas, A; Gates, C; Robbins, P; Wernig, A; Huard, J. Clonal isolation of muscle-derived cells capable of enhancing muscle regeneration and bone healing. *J. Cell Biol.,* 2000, 150, 1085-1100.

Lee, JY; Musgrave, D; Pelinkovic, D; Fukushima, K; Cummins, J; Usas, A; Robbins, P; Fu, FH; Huard, J. Effect of bone morphogenetic protein-2-expressing muscle-derived cells on healing of critical-sized bone defects in mice. *J. Bone and Joint Surg*., 2001,83A, 1032-1039.

Levander, G. A study of bone regeneration. *Surg. Gyn. Obst.,* 1938, 67, 705-714.

Levander, G. Tissue induction. *Nature,* 1945, 155, 148-149.

Levander, G; Willstaedt, H. Alcohol-soluble osteogenetic substance from bone marrow. *Nature,* 1946, 157, 587.

Luyten, FP; Cunningham, NS; Ma, S; Muthukumaran, N; Hammonds, RG; Nevins, WB; Woods, WI; Reddi, AH. Purification and partial amino acid sequence of osteogenin, a protein initiating bone differentiation. *J. Biol. Chem.*, 1989, 264, 13377-13380.

McBrearty, BA; Clark, LD, Zhang, XM; Blakenhorn, EP; Heber-Katz, E. Genetic analysis of a mammalian wound-healing trait. *Proc. Natl. Acad. Sci. USA,* 1998, 95, 11792-11797.

Medici, D; Shore, E; Lounev, VY; Kaplan, FS; Kalluri, R; Olsen, BR. Conversion of vascular endothelial cells into multipotent stem-like cells. *Nature Medicine*, 2010, 16, 1400-1406.

Minasi, MG; Riminucci, M; De Angelis, L; Borello, U; Berarducc,i B; Innocenzi, A; Caprioli, A; Sirabella, D; Baiocchi, M; De Maria, R; Boratto, R; Jaffredo, T; Broccoli, V; Bianco, P; Cossu, G. The meso-angioblast: a multipotent, self-renewing cell that originates from the dorsal aorta and differentiates into most mesodermal tissues. *Development 2002*, 129, 2773-2783.

Miyazawa, K; Shinozaki, M; Hara, T; Furuya, T; Miyazono, K. Two major Smad pathways in TGF-beta superfamily signalling. *Genes Cells*, 2002, 12, 1191-1204.

Nagatomo, K; Komaki, M; Sekiya, I; Sakaguchi, Y; Noguchi, K; Oda, S; Muneta, T; Ishikawa, I.Stem cell properties of human periodontal ligament cells. *J. Periodont. Res.,* 2006, 41, 303-310.

Peault, B; Rudnicki, M; Torrente, Y; Cossu, G; Tremblay, JP; Partridge, T; Gussoni, E; Kunkel, LM; Huard, J. Stem and progenitor cells in skeletal muscle development, maintenance, and therapy. *Mol. Ther.*, 2007, 15, 867-877.

Reddi, AH. Extracellular matrix and development, In: K.A. Piez and A.H.Reddi (Eds.) *Extracellular Matrix Biochemistry New York:Elsevier*, 1984, 375-412.

Reddi, AH. Symbiosis of biotechnology and biomaterials: applications in tissue engineering of bone and cartilage. J Cell Biochem, 1994, 56, 192-195.

Reddi, AH. Bone morphogenesis and modeling: soluble signals sculpt osteosomes in the solid state. *Cell,* 1997, 89, 159-161.

Reddi, AH. Role of morphogenetic proteins in skeletal tissue engineering and regeneration. *Nat. Biotechnol.*, 1998, 16, 247-252.

Reddi AH. Morphogenesis and tissue engineering of bone and cartilage: Inductive signals, stem cells, and biomimetic biomaterials. *Tissue Eng.,* 2000, 6, 351-359.

Reddi, AH; Huggins, C. Biochemical sequences in the transformation of normal fibroblasts in adolescent rats. *Proc. Natl. Acad. Sci. USA*, 1972, 69, 1601-1605.

Ripamonti, U. Bone induction by recombinant human osteogenic protein-1 (hOP-1, BMP-7) in the primate papio ursinus with expression of mRNA of gene products of the TGF-beta superfamily. *J. Cell Mol. Med.,* 2005, 9, 911-928.

Ripamonti, U. Soluble osteogenic molecular signals and the induction of bone formation. *Biomaterials,* 2006, 27, 807-822.

Ripamonti, U. The Marshall Urist Awarded Lecture. Bone: Formation by autoinduction. In: Vukicevic S, Reddi AH, eds. Proceedings of the 6th International Conference on BMP's. Dubrovnick, 11-16 October 2006a: p.1.

Ripamonti, U. Therapeutic Tissue Engineering – Fact or Fiction? Pre-clinical animal models and translational research in Homo sapiens: Fact or Fiction? Proceedings of the International Association of Anatomists, Cape Town, South Africa. 2009 Available at URL http://www.ifaa.net/IFAA2009.pdf.

Ripamonti, U. Soluble and insoluble signals sculpt osteogenesis in angiogenesis. *World J. Biol. Chem.,* 2010, 1, 109-132.

Ripamonti U; Bosch C; van den Heever B; Duneas N; Melsen B; Ebner R. Limited chondro-osteogenesis by recombinant human transforming growth factor-β1 in calvarial defects of adult baboons (Papio ursinus). *J. Bone Miner. Res.*, 1996, 11, 938-945.

Ripamonti, U; Crooks, J; Kirkbride, AN. Sintered porous hydroxyapatites with intrinsic osteoinductive activity: Geometric induction of bone formation. *South Afr. J. Sci.,* 1999, 335-343.

Ripamonti, U; Crooks, J; Matsaba, T; Tasker, J. Induction of endochondral bone formation by recombinant human transforming growth factor β2 in the baboon (Papio ursinus). *Growth Factor*, 2000, 17, 269-285.

Ripamonti, U; Crooks, J; Teare, J; Petit, J-C; Rueger, DC. Periodontal tissue regeneration by recombinant human osteogenic protein-1 in periodontally-induced furcation defects of the primate Papio ursinus. *South Afr. J. Sci.,* 2002, 98, 361-368.

Ripamonti, U; Duneas, N; van den Heever, B; Bosch, C; Crooks, J. Recombinant transforming growth factor- β1 induces endochondral bone in the baboon and synergizes with recombinant osteogenic protein-1 (bone morphogenetic protein-7) to initiate rapid bone formation. *J. Bone Miner. Res.,* 1997, 12, 1584-1595.

Ripamonti, U; Ferretti, C. Mandibular reconstruction using naturally-derived bone morphogenetic proteins: A clinical trial report. In: T.S. Lindholm (Ed.), Advances in skeletal reconstruction using bone morphogenetic proteins. *World Scientific Publ., Co. Singapore*, 2002, 277-289.

Ripamonti, U; Ferretti C. Grand challenges for craniomandibulofacial reconstruction by human recombinant transforming growth factor-β3. In: R.S. Tuan, F. Guilak, and A. Atala (Eds.), Proceedings of the Keystone Symposia on Regenerative Tissue Engineering and Transplantation 1-6 April 2012, Breckenridge, Colorado, Available at: http://www.keystonesymposia.org.

Ripamonti, U; Ferretti, C; Heliotis, M. Soluble and insoluble signals and the induction of bone formation: Molecular therapeutics recapitulating development. *J. Anat.*, 2006, 209, 447-468.

Ripamonti, U; Ferretti, C; Teare, J; Blann, L. Transforming growth factor-β isoforms and the induction of bone formation: Implications for reconstructive craniofacial surgery. *J. Craniofac. Surg*., 2009c, 20, 1544-1555.

Ripamonti, U; Heliotis, M; Ferretti, C. Bone morphogenetic proteins and the induction of bone formation: From laboratory to patients. *Oral Maxillofac. Surg. Clin. North Am.,* 2007, 19, 575-589, vii.

Ripamonti, U; Heliotis, M; van den Heever, B; Reddi, AH. Bone morphogenetic proteins induce periodontal regeneration in the baboon (Papio ursinus). *J. Periodontal. Res*., 1994, 29, 439-445.

Ripamonti, U; Klar, RM; Renton, LF; Ferretti, C. Synergistic induction of bone formation by hOP-1, hTGF-β3 and inhibition by zoledronate in macroporous coral-derived hydroxyapatites. *Biomaterials,* 2010, 31, 6400-6410.

Ripamonti, U; Ma, SS; Reddi, AH. Induction of bone in composites of osteogenin and porous hydroxyapatite in baboons. *Plast. Reconstr. Surg.*, 1992, 89, 731-739; discussion 740.

Ripamonti, U; Parak, R; Petit, JC. Induction of cementogenesis and periodontal ligament regeneration by recombinant human transforming growth factor-β3 in Matrigel with rectus abdominis responding cells. *J. Periodontal. Res.*, 2009a, 44, 81-87.

Ripamonti U; Petit J-C. Bone morphogenetic proteins, cementogenesis, myoblastic stem cells and the induction of periodontal tissue regeneration. *Cytokine. Growth Factor Rev.,* 2009, 20, 489-499.

Ripamonti, U; Petit, JC; Teare, J. Cementogenesis and the induction of periodontal tissue regeneration by the osteogenic proteins of the transforming growth factor-βsuperfamily. *J. Periodontal. Res.*, 2009b, 44, 141-152.

Ripamonti, U; Ramoshebi, LN; Matsaba, T; Tasker, J; Crooks, J; Teare, J. Bone induction by BMPs/OPs and related family members in primates. *J. Bone Joint Surg. Am.,* 2001, 83-A Suppl 1(Pt 2), S116-127.

Ripamonti U; Ramoshebi LN; Teare J; Renton L; Ferretti C. The induction of endochondral bone formation by transforming growth factor-β3: Experimental studies in the non-human primate Papio ursinus. *J. Cell Mol. Med.*, 2008, 12, 1029-1048.

Ripamonti U; Reddi AH. Periodontal regeneration: potential role of bone morphogenetic proteins. *J. Periodont. Res.*, 1994, 29, 225-235.

Ripamonti, U; Reddi AH. Bone morphogenetic proteins: Applications in plastic and reconstructive surgery. *Advances in Plastic and Reconstructive Surgery,* 1995, 11, 47-65.

Ripamonti, U; Reddi AH. Tissue engineering, morphogenesis, and regeneration of the periodontal tissues by bone morphogenetic proteins. *Crit. Rev. Oral Biol. Med.,* 1997, 8, 154-163.

Ripamonti, U; Ramoshebi, LN; Patton, J; Matsaba, T; Teare, J; Renton, L. Soluble signals and insoluble substrata: novel molecular cues instructing the induction of bone In: E.J. Massaro, J.M. Rogers (Eds.) The Skeleton Totowa: Human Press, 2004, 217-227.

Ripamonti, U; Richter, PW; Nilen, RW; Renton, L. The induction of bone formation by smart biphasic hydroxyapatite tricalcium phosphate biomimetic matrices in the non-human primate Papio ursinus. *J. Cell Mol. Med.*, 2008a, 12, 2609-2621.

Ripamonti, U., Richter, PW; Thomas, ME. Self-inducing shape memory geometric cues embedded within smart hydroxyapatite-based biomimetic matrices. *Plast. Reconstr. Surg.*, 2007a, 120, 1796-1807.

Ripamonti, U; Roden, LC. Induction of bone formation by transforming growth factor-β2 in the non-human primate Papio ursinus and its modulation by skeletal muscle responding stem cells. *Cell Prolif.*, 2010, 43, 207-218.

Ripamonti, U; Roden, LC, Renton, LF. Osteoinductive hydroxyapatite-coated titanium implants. Biomaterials, 2012, doi S0142-9612(12)00096-8 [pii] 10.1016/j.*biomaterials.*,2012.01.050.

Ripamonti, U; Teare, J; Ferretti, C. A macroporous bioreactor super activated by the recombinant human transforming growth factor-β3. *Frontiers in Physiology,* 2012, 3, 1-18.

Ripamonti, U; van den Heever, B; Sampath, TK; Tucker, MM; Rueger, DC; Reddi, AH. Complete regeneration of bone in the baboon by recombinant human osteogenic protein-1 (hOP-1, bone morphogenetic protein-7). *Growth Factors*, 1996, 13, 273-289.

Ripamonti, U; van den Heever, B; Van Wyk, J. Expression of the osteogenic phenotype in porous hydroxyapatite implanted extraskeletally in baboons. *Matrix,* 1993, 13, 491-502.

Roberts, AB; Sporn, MB; Assoian, RK; Smith, JM; Roche, NS; Wakefield, LM; Heine,UI; Liotta, LA; Falanga, V; Kehrl, JH; Fauci, AS. Transforming growth factor type β: rapid induction of fibrosis and angiogenesis in vivo and stimulation of collagen formation in vitro. *Proc. Natl. Acad. Sci. USA,* 1986, 83, 4167–4171.

Sacerdotti, C; Frattin, G. Sulla produzione eteroplastica dell'osso. *R. Accad. Med. Torino.,* 1901, 825-836.

Sampath, TK; Maliakal, JC; Hauschka, PV; Jones, WK; Sasak, H; Tucker, RF; White, KH; Coughlin, JE; Tucker, MM; Pang, RH; et al. Recombinant human osteogenic protein-1 (hOP-1) induces new bone formation in vivo with a specific activity comparable with natural bovine osteogenic protein and stimulates osteoblast proliferation and differentiation in vitro. *J. Biol. Chem.*, 1992, 267, 20352-20362.

Sampath, TK; Reddi, AH. Dissociative extraction and reconstitution of extracellular matrix components involved in local bone differentiation. *Proc. Natl. Acad. Sci. USA,* 1981,78, 7599-7603.

Sampath, TK; Reddi, AH. Homology of bone-inductive proteins from human, monkey,bovine, and rat extracellular matrix. *Proc. Natl. Acad. Sci. USA*, 1983, 80, 6591-6595.

Sampath, TK; Reddi, AH. Distribution of bone inductive proteins in mineralized and demineralized extracellular matrix. *Biochem. Biophys. Res. Commun.,* 1984, 119, 949-954.

Schwartz, SM; Benditt EP. Aortic endothelial cell replication I. Effects of age and hypertension in the rat. *Circ. Res.*, 1977, 41, 248-255.

Senn, N. On the healing of aseptic bone cavities by implantation of antiseptic decalcified bone. *AM. J. Med. Sci.,* 1889, 98, 219–243.

Sigurdsson, TJ; Lee, MB; Kubota, K; Turek, TJ; Woney, JM; Wikesjo, UME. Periodontal repair in dogs: recombinant human bone morphogenetic protein-2 significantly enhances periodontal regeneration. *J. Periodontol.*, 66, 131-138.

Stevens, MM; Marini, RP; Schaefer, D; Aronson, J; Langer, R; Shastri, VP. In vivo engineering of organs: the bone bioreactor. Proc. Natl. Acad. Sci. USA, 2005, 102, 11450-11455.

Teare, JA; Ramoshebi, LN; Ripamonti U. Periodontal tissue regeneration by recombinant human transforming growth factor-β3 in Papio ursinus. *J. Periodontal. Res.*, 2008, 43, 1-8.

Turing, AM. The chemical basis of morphogenesis. *Philos. Trans. Roy. Soc. Lond.*, 1952, 237, 37-41.

Urist, MR. Bone: Formation by autoinduction. *Science,* 1965, 150, 893-899.

Urist, MR; Dowell, TA; Hay PH; Strates, BS. Inductive substrates for bone formation. *Clin. Orthop. Relat. Res.*, 1968, 59, 59-96.

Urist, MR; Silverman, BF; Buring, K; Dubuc, FL; Rosenberg, JM. The bone induction principle. *Clin. Orthop. Relat. Res.*, 1967, 53, 243-283.

Urist, MR; Mikulski, A; Lietze, A. Solubilized and insolubilized bone morphogenetic protein. *Proc. Natl. Acad. Sci. USA*, 1979, 76, 1828-1832.

Urist, MR; Strates, BS. Bone morphogenetic protein. *J. Dent. Res.*, 1971, 50, 1392-1406.

Ūsas, A; Huard, J. Muscle-derived stem cells for tissue engineering and regenerative therapy. *Biomaterials*, 2007, 28, 5401-5406.

Ūsas A, Mačiulaitis J, Mačiulaitis R, Jakubonienė N, Milašius A and Huard J. Skeletal muscle-derived stem cells: Implications for cell-mediated therapies. *Medicina (Kaunas) 2011*, 47, 469-479.

Wang, EA; Rosen, V; Cordes, P; Hewick, RM; Kriz, MJ; Luxenberg, DP; Sibley, BS; Wozney, J M. Purification and characterization of other distinct bone-inducing factors. *Proc. Natl. Acad. Sci. USA,* 1988, 85, 9484-9488.

Wang, EA; Rosen, V; D'Alessandro, JS; Bauduy, M; Cordes, P; Harada, T; Israel, D1; Hewick, RM; Kerns, KM; Lapan, P; Luxenberg, DP; McQuaid, D; Moutsatso, IK; Nove, J; Wozney, JM. Recombinant human bone morphogenetic protein induces bone formation. *Proc. Nat. Acad. Sci. USA,* 1990, 87, 2220-2224.

Williams, DF. Tissue Engineering: the multidisciplinary epitome of hope and despair. eds. Paton, R., Macnamare, L. (Amsterdam: Elsevier BV), 2006, 483-524.

Wozney, JM; Rosen, V; Celeste, AJ; Mitsock, LM; Whitters, MJ; Kriz, RW; Hewick, RM; Wang, EA. Novel regulators of bone formation: molecular clones and activities. *Science,* 1988, 242, 1528-1534.

Young, HE; Duplaa, C; Young, TM; Foyd, JA; Reeves, ML; Davis, KH; Mancini, GJ; Eaton, ME; Hill, JD, Thomas, K; Austin, T; Edwards, C; Cuzzourt, J; Parikh, A; Groom, J; Hudson, J; Black, AC Jnr. Clonogenic Analysis Reveals Reserve Stem Cells in Postnatal Mammals: I.

Pluripotent Mesenchymal Stem Cells. *The Anatomical Record 2001*, 263, 350-360.

Zheng, B; Cao, B; Crisan, M; Sun, B; Li, G; Logar, A; Yap, S; Pollett, JB; Drowley, L; Cassino, T; Gharaibeh, B; Deasy, BM; Huard, J; Peault, B. Prospective identification of myogenic endothelial cells in human skeletal muscle. *Nat. Biotechnol. 2007*, 25, 1025-1034.

Zhao, ZM; Berry, JE; Somermann, MJ. Bone morphogenetic protein-2 inhibits differentiation and mineralization of cementoblasts in vitro. *J. Dent. Res.*, 2003, 82, 23-27.

In: Muscle Cells
Editor: Benigno Pezzo

ISBN: 978-1-62417-233-5


*Chapter 3*

# Pathogenesis of Inguinal Hernia and Hydrocele: The Role of Muscle Cells on the Processus Vaginalis

***Vassilios Mouravas* and Dimitrios Sfoungaris***
Second Department of Pediatric Surgery,
Aristotle University of Thessaloniki,
General Hospital Papageorgiou, Thessaloniki, Greece

## Abstract

Congenital inguinal hernia (IH) and hydrocele are among the commonest pathologies affecting children and both are caused by the incomplete obliteration of the processus vaginalis (PV) which normaly obliterates near the end of the gestational period or sortly after.

A number of factors, endocrine, neurophysiologic, cytologic, regulate PV development. These regulatory factors are not mutually exclusive in their action and we think that an experimental or observational finding that may affect the fate of the PV does not necessarily invalidate a seemingly contradictory theory based on other findings.

* Address: Tzavela 8 str, 55535, Thessaloniki, Greece. E-mail:vmourav1@otenet.gr

The normal process of PV obliteration is considered, by some authors, to include a stage of dedifferentiation of smooth muscle cells (SMCs) that are found on the PV, and their eventual apoptosis. Histological studies reveal the existence of SMCs on the wall of unobliterated PV. Sympathetic and parasympathetic nerve action, which in its turn is affected by hormones, is probably involved to produce or to halt such a result.

In this particular study we review the literature on these biologic mechanisms, including our own contribution which is the following: By using immuno-histochemical studies we examined the cytoskeletal proteins of SMCs present in the PV of patients with IH and hydrocele and drew conclusions on the degree of SMC dedifferentiation. Sacs from patients with IH and especially from male IH, have fully differentiated SMCs while sacs obtained from hydroceles are in an intermediate state of dedifferentiation. Our findings are suggestive that in cases of IHs the SMCs on the wall of the hernia sac do not follow the natural way of dedifferentiation and apoptosis, and only partly do so in cases of hydrocele. This may be the reason for the varying degree of incomplete obliteration of the PV in these cases.

## INTRODUCTION

The anatomy of the processus vaginalis and the neighbouring structures of the inguinal canal have been extensively studied and described through all the stages of development, from fetus to old age. The inguinal region, especially in the male, has attracted the attention of practitians and scientists since the antiquity because it is the site of very common pathologies, affecting humans of all ages. The fact that at this region of the body, some internal organs cross the abdominal barrier and place themselves more superficially, apparently causing disturbances during their course, offered a privileged terrain for anatomists, histologists and embryologists.

A number of biologic mechanisms, generally related to gonadal development and migration, have been postulated to initiate, promote, stop or cause regression in the course of developmental transformations. Even though considerable information has been accumulated on the subject, not necessarily contradictory to one eachother, several important controversies do exist.

From a clinicoanatomic point of view, several mechanisms of PV obliteration have been postulated: (a) a progressive fibrous closure that starts above the epididymis and proceeds in a cephalad direction, (b) compression of the PV by surrounding tissue until it turns into a narrow tubular structure which finaly obliterates by fibrosis and (c) segmentation of the funicular portion followed by variable fibrous obliteration [1].

Indirect IH, communicating hydrocele and spermatic cord cyst, all have a common feature: the processus vaginalis is not obliterated, as it should, at birth or shortly after. Another common pathology of the region, cryptorchidism, is also accompanied by an unobliterated PV which does not give rise to a hernia [1]. The use of immunohistochemical studies makes it possible to trace the fate of individual cells that develop in the PV, correlate their developmental stage with pathology, and draw conclusions about the mechanisms involved in the process of PV obliteration.

In this paper we review the current state of information on the subject, based on other authors' research as well as on ours.

Since he PV is in direct relation to the gumbernaculum testis (GT) during the various stages of development, its morphology can only be studied in connection with it and with testicular descent. In the same way, its obliteration can be understood only in the context of the development of the related tissues and the biological processes that affect them.

## EMBRYOLOGY

The PV constitutes a blind process of the peritoneum developing inside the gumbernacular mesenchyme, whose early development is observed in stage 14 CC (5-7 mm CRL) embryos as the caudal genito-inguinal ligament. It connects the lower pole of the gonad and epididymis to the future site of the inner inguinal ring and inguinal canal.

During the embryonic stage 20-23 CC (21-30 mm CRL), three parts of the GT (abdominal, interstitial and subcutaneous) are distinguished. The PV appears with its dorsal layer attached to the ventral side of the GT. In a later stage (32-55 mm CRL) an enormous increase in length and volume of the GT takes place along with an enlargement of the PV [2]. The processus vaginalis is developing inside the GT and divides it into three parts.

The outer rim of gumbernacular mesenchyme is where the cremaster muscle forms. The median part represents the invaginating PV, and the innermost is the central column or cord that attaches to the caudal epididymis and testis. Caudal to the PV is the solid tip of the GT which contains abundant undifferentiated mesenchyme and glucopolysaccharides. It is quite bulky, being about the same size as the testis [2].

The GT remains bulky and gelatinous until after migration through the IC is complete. This migration phase requires the GT to change from a relatively inert, static structure ending in the inguinal muscles into an elongating, migrating organ that extends across the pubis and into the scrotum in the perineum [3].

Between 10-15 weeks of development the testis remains near the future inguinal canal. At arround 25-28 weeks the testis descends rapidly through the inguinal canal, which has just formed, and then migrates across the pubic region and down into the scrotum, arriving there at about 35–40 weeks The distance required for the GT to transverse is considerable, being more than 4cm in many fetuses, when the GT itself is only 1cm in diameter [3]. As the GT contains the PV which is a peritoneal diverticulum, intra-abdominal pressure is transmitted into it and contributes in inguinoscrotal descent [4]. The high occurrence of IH and hydroceles after ventriculoperitoneal shunt insertion supports the role of raised intraabdominal pressure as an etiological factor for these conditions [5].

By observing these extensive morphological changes of the PV and the GT, researchers reached diverging conclusions as to which structures change size and direction primarily, i.e. which of them are the driving force that drift neighbouring structures, and which change secondarily, i.e. they passively follow the primary changes. The vast majority of research dissections performed in the scope of clarifying these processes are performed on animals that do not necessarily develop the same way as humans do. This is a reason for a clear picture to be at some extent missing on this stage of testicular migration [6].

Some researchers consider that the PV elongates passively, responding to the shortening of the GT and the descent of the testis towards the scrotum [7], [8]. Others attribute a more energetic role at the PV. According to them, the PV takes an active part in its own elongation, opening up a way by forming the inguinal canal, in order to facilitate the testicular descent [9]. There is evidence suggesting that the GT acquires specific growth properties, similar to an embryonic limb bud, enabling both the processus vaginalis and the cremaster muscle to grow maximally from their distal end [3].

Animal experiments and observations have equally shown an active role for the GT and the PV. In cat cubs, the PV undergoes extensive cellular proliferation at its distal end showing an active elongation that drives the testis to the scrotum [10].

In the mouse, the PV seems to be derived from the surface of the urogenital ridge, separate from the remaining parietal peritoneum suggesting that the PV has evolved to aid testicular descent in this species, rather than being an inert diverticulum of the parietal peritoneum [11]. In the rat, electron microscopy revealed that PV developed, while the conus of the GT disappeared, after which the testis moved out of the abdominal cavity and entered the PV [6].

## HISTOLOGY OF OBLITERATED PV AND HERNIA SACS

Histological findings in the clinical conditions of IH and hydrocele are as follows (Table 1):

**Table 1. Histological Findings in IH and Hydroceles**

| | Obliterated PV | IH boys | IH girls | Hydrocele |
|---|---|---|---|---|
| Smooth muscle layer | No | Yes bundles | Yes bundles | Few patchy |
| Myofibroblasts | No | Yes | Yes | No |
| Striated muscle | No | No | No | No |

*IH sacs of boys and girls:* The innermost layer is consisted of mesothelium, in continuation to the peritoneum. This is surrounded by a supporting layer of loose connective tissue containing blood vessels and peripheral nerves. An outer layer is formed by SMCs which are organized in bundles. Myofibroblasts were observed [12].

*Hydrocele sacs:* They exhibit the same features as the IH sacs, except that there are no recognizable SMC bundles. Instead SMCs are to be found dispersed in the loose connective tissue layer [12].

Histologic examination of the obliterated PV demonstrates an inner layer of mesothelium and a supporting layer of loose connective tissue containing blood vessels and peripheral nerves [13, 14].

## INTRACELLULAR MATRIX

As mentioned above, the PV and GT undergo extensive changes within their matrix components. The GT during the early gestational period when the testes are abdominal in position, has a rather loose and hyaline extracellular matrix with few collagen fibers. Fibroblasts are the most abundant cells, homogeneously distributed in the whole GT. From 20 to 24 weeks of gestation the amount of collagen increases and the extracellular matrix becomes moderately dense. Cell density then decreases sharply with age, and by 29 weeks when the testes are in the scrotum, the intercellular space is noticeably larger. At 28 weeks, the extracellular matrix is dense with abundant collagen fibers and by 29 weeks it is even denser resembling a mature scar. By the 28 week elastic fibers are revealed, preferentially located at the distal end of the GT where they gradually replace striated muscle cell bundles [15].

Abnormal collagen which appears in the matrix of the PV, either due to local factors or to general metabolic deviation, is thought to play an important role in the development of IH. A marked attenuation of the transversalis fascia and a significant reduction in the thickness of connective tissue was demonstrated in the area of the internal ring on the clinically normal side of patients with IH [16].

The detection of an impaired collagen balance in the tissue as well as in cultured fibroblasts contributes to a decreased tensile strength and mechanical stability of the connective tissue. These findings support the hypothesis of a systemic disease rather than a mere local mechanical defect as etiologic factors for hernia formation [17].

The altered ratio of the collagen subtypes can result either by a modified synthesis or by an imbalanced breakdown. A significant increase in type III procollagen synthesis in fibroblasts from patients with IH has been detected, which may result in reduced collagen fibril assembly in the abdominal wall, eventually leading to herniation. It is not yet clear what genetic factors are responsible for the increase in type III collagen synthesis [18]. The cleavage of matrix collagen is regulated by the activity of the matrix metallo-proteinases (MMPs).

Among them are the principal matrix enzymes cleaving fibrillar type I, II and III collagen. Serum levels of MMP-2 were significantly increased in all the hernia patients as compared to controls, in cases of direct and indirect hernia (congenital type) hernia [17].

Other investigators have failed to confirm several of the abovementioned results. The expression pattern of type I and III collagen did not differ among sacs obtained from patients with IH, hydrocele and undescended testis when compared with that of controls. However, strong expression of type III collagen was observed in the hernial sacs of right-sided male IH compared with left side [19].

## MESOTHELIUM

Fusion of the mesothelial opposing layers and tissue remodelling is observed in PV obliteration. Experimental evidence has suggested several pathways leading to obliteration.

Calcitonin gene-related peptide (CGRP), which is released from the genitofemoral nerve, may trigger fusion of the patent processus vaginalis in children with IH. Cultured epithelial cells derived from the patent processus vaginalis were analysed by in vitro culture in the presence of several factors. Epithelial and mesenchymal markers underwent either down-regulation or up-regulation as epithelial cell sheets broke apart and individual cells started to migrate. Hepatocyte growth factor (HGF) produced transformation of hernial sac epithelial cells, whereas CGRP could act indirectly via HGF, which, in turn, promotes fusion of the processus vaginalis. The author hypothesizes that in the future, a nonsurgical treatment of IH in children might be possible by the local administration of agents which promote fusion. [20]. HGF was found to induce fusion of PV and may be involved as an intermediate molecule in the fusion cascade [21].

Evidence gained from alveolar epithelial cell death suggests the myofibroblasts play an important role in the programmed cell death of the mesothelial layer. Apoptosis has been observed to primarily take place in alveolar cells adjacent to myofibroblasts. Myofibroblasts are suggested to induce apoptosis in alveolar cells by producing some soluble inducers. By analogy, myofibroblasts may also take part in the disappearance of the mesothelial layer of the processus vaginalis [22].

## MUSCLE CELLS

Striated muscle cells have been reported to appear by the eighth week of gestation within the gubernaculum [9] but this finding was not confirmed by others. With the use of immunohistochemistry, which enables a more accurate diagnosis, muscle structures are revealed in the human GT at twelve weeks of gestation. These structures, represent fragmented myotubes positive for both human muscle actin and desmin. On the contrary, Myo-D was not expressed. According to Tanyel [14], the striated muscles which do not express Myo-D represent the projecting muscles of the abdominal wall, which will eventually cease to exist. Myo-D was equally not expressed in the vascular SMCs, which were positive for only human muscle actin.

Actin and desmin expressing striated muscles ceased to exist until 22 weeks of gestation. Both vascular SMCs and cremaster striated muscle (CStM) expressed Myo-D during the 22nd and 23rd weeks. This synchronous detection of Myo-D in both SM and CStM suggests that CStM may have transdifferentiated from the vascular SM. CStM additionally expressed alpha-smooth muscle actin (aSMA). Other researchers have found SMCs restricted to the walls of blood vessels. Striated muscle cells were detected at the scrotal end of the GT, appearing as isolated and scattered bundles running in various directions. Like fibroblasts, their number also decreased with age [15]. Detection of myofibroblasts in the 22nd week was followed by differentiation towards SMCs which appear by the 27th week in the gubernacula of male fetuses. The same alterations were encountered among the female fetuses at later time [23].

According to these findings, under normal conditions, SMCs are only transiently present in the GT and PV. They appear starting differentiation in 12-19 weeks fetuses, are still apparent in fetuses 20-25 weeks, and they undergo degeneration and disappear after the testis reaches the scrotum. IH sacs of boys and girls contain smooth muscle, while obliterated PV do not, and smooth muscle bundles are only sparsely present in sacs associated with hydrocele [22, 24].

While sacs from boys contained only smooth muscle, sacs from girls demonstrated also striated muscle. Myofibroblasts may have originated from the smooth muscle, and reflect the attempts at obliteration of PV [24, 25].

Ascertained through electron microscopy as well, myofibroblasts were commonly encountered in sacs associated with IH and smooth muscle was invariably present in sacs that contained myofibroblasts. Myofibroblasts are found in association with smooth muscle and thus, such cells within the sac walls seem to originate from the smooth muscle, reflecting the process of dedifferentiation. This dedifferentiated state may represent attempted apoptosis, which usually causes the disappearance of the smooth muscle and allows obliteration of the processus vaginalis [22].

All this evidence suggests that the persistence of smooth muscle hinders the obliteration of the processus vaginalis and influences the clinical outcome [22, 24]. Apoptotic nuclei have been detected within the vascular structures and mesothelium. However, none of the samples from different diagnostic sources have revealed any apoptotic nuclei within the smooth muscle component. The failed apoptosis of smooth muscle may have a role in the persistence of PV [26].

During the process of ontogenesis a large number of cells cease to exist after a certain period, when they have accomplished their purpose. They do so by the mechanism of programmed cell death, also known as apoptosis. The end point of the mechanism is the intracellular formation of caspases, a group of proteases that disintegrate cellular proteins. These are formed when the mitochondrial membranes are severed and Cytochrome C enters the cytosol. Mitochondrial integrity can be affected by a variety of mediators produced after the activation of the Bax-BCL2 and the Fas/Ligand systems. It is also affected by the depletion of the endoplasmic reticulum from Ca++ and the concomitant cytosolic and mitochondrial increased Ca++. Both sympathetic and parasympathetic innervation can influence this course, and this in turn depends on exposure to androgen [27].

## Neuronal and Humoral Factors

Two different theories have been proposed implicating neuronal and humoral factors in GT and PV development. The genitofemoral nerve (GFN) hypothesis postulates that gubernacular migration from the inguinal region to the bottom of the scrotum is controlled by a neurotransmitter released from the GFN [28]. The genital branch of the GFN enters the inguinal canal through the deep inguinal ring and reaches the scrotum to supply the GT and the coverings of the spermatic cord [29] and releases CGRP from its sensory nerve endings [30]. Androgens appear to act both directly on the gubernaculum and indirectly via the GFN. The exact site of androgen effects on the genitofemoral nerve are not known but sexual dimorphism has been demonstrated in the cell bodies of the dorsal root ganglion. The primary hormone regulating transabdominal descent is insulin-like hormone 3 (Insl3), which is secreted by the Leydig cells and stimulates the swelling reaction by a receptor on the gubernaculums [31].

Central catecholaminergic activity affects both sympathetic tonus and GnRH secretion which promotes androgen secretion. Increased sympathetic tonus, acting via beta-adrenergic receptors and intracellular cAMP, induces trophic influences upon SMCs.

Since sympathetic tonus is androgen-dependent, and since smooth muscle also responds to androgens (a well known effect on prostatic muscle), at least two pathways exist to exert androgen effects upon smooth muscle. These effects do not allow SMCs to initiate the dedifferentiation pathway, allowing their persistence on the wall of the PV, halting the obliteration, and predisposing to IH [12, 24, 27].

A transient decrease in sympathetic and increase in parasympathetic tonus during a critical time and with a critical intensity is the physiologic event and the requirement for the obliteration of the PV. If the decrease in sympathetic tonus is enough to deplete the calcium stores and to increase the cytosolic calcium of the SMCs, but not profound enough or does not sustain enough to increase the Bax and Fas levels, it results in a lesser degree of apoptosis and the possible result of a hydrocele [27].

The sympathetic tonus is sexually dimorphic and it is less in females than in males. Its suppression below a critical level is therefore easier and more probable and this fact may contribute to the lower incidence of IH in females [25, 27].

## Immunohistochemical Study of the Hernia Sac in Children with IH and Hydrocele

We investigated the diversity and differentiation of smooth muscle phenotypes in sacs associated with inguinal hernia and hydrocele through the expression of aSMA, h-caldesmon, desmin, and vimentin. The examined PVs originated from boys with IH (n = 23), girls with IH (n = 8), and boys with hydrocele (n = 10). Peritoneal samples (male, 4; female, 3) and obliterated PV (male, 3) obtained from age-matched patients served as controls. The samples were treated accordingly and evaluated immunohistochemically using monoclonal antibodies against the abovementioned proteins. No presence of SMCs was evident in control samples. The expression of aSMA, desmin, and h-caldesmon did not differ among sacs obtained from patients with inguinal hernia and hydrocele. However, a strong expression of vimentin in SMCs of hydrocele sacs, in comparison to sacs from male patients with inguinal hernia was observed (Table 2)

Our findings, as well as those by other researchers, affirm the presence of SMCs on unobliterated PV (Figure 1) (Figure 2). These SMCs present in two distinct phenotypes that are identified by their ultrastructural equipment and by the type of cytoskeletal proteins they contain. Immature SMCs that present a synthetic phenotype are characterized by well-developed synthetic organelles, especially Golgi apparatus, and reduced contractile myofilaments.

**Table 2. Expressed Proteins in IH, Hydrocele, and Controls**

| Markers | Control | Inguinal hernia | Hydrocele |
|---|---|---|---|
| aSMA, hcaldesmon and desmin | No | Yes | Yes |
| vimentin | No | 11/23 boys<br>6/8 girls | 10/10 |

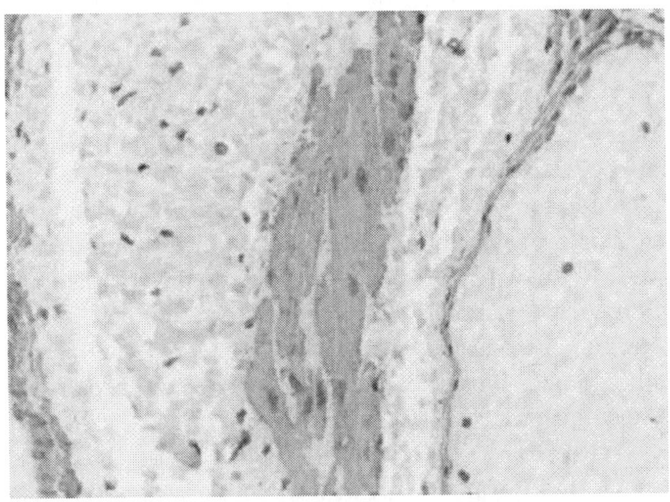

Figure 1. a-SMA positive bundles in a boy 3 years old with IH, X400.

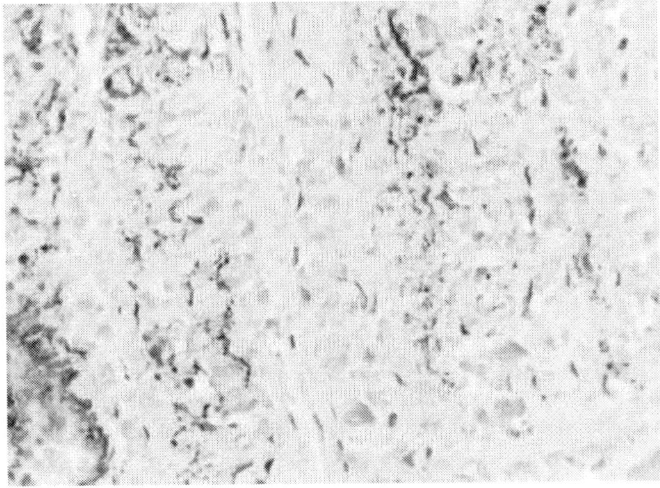

Figure 2. Few a-SMA positive SMCs not organized in bundles, in the PV of a 5 years old boy with hydrocele.

These cells express vimentin, as an intermediate filament. Vimentin is recognized, among other intermediate filaments, as a marker of undifferentiated SMCs.

Mature SMCs, that present a highly contractile fully differentiated phenotype, are characterized by a well-developed system of contractile myofilaments, instead of synthetic organelles. These cells express a low amount of vimentin but a high amount of desmin, aSMA and h-caldesmon (Table 2).

These proteins have been used as markers of differentiated SMCs. During development, the process of differentiation/maturation of SMCs from the synthetic to the contractile phenotype is accompanied by ultrastructural changes that correlate with a progressively increased expression of desmin, actin and h-caldesmon, and decreased expression of vimentin. It has been observed that this process, under normal conditions during the development, reverses and SMCs that have reached maturity de-differentiate and achieve a more undifferentiated state. This has been, for example, observed in cell cultures or during atherogenesis, where SMCs undergo dedifferentiation from the contractile to the synthetic phenotype. Neuronal and humoral factors thought to affect this process are mentioned earlier in this text.

**Table 3. Dedifferen-tiation process and associated markers**

| SMC status | Mature | Immature | Apoptosis |
|---|---|---|---|
| | → | | |
| Markers | Desmin +++<br>aSMA +++<br>h-caldesmone +++<br>Vimentine + | Desmin -<br>aSMA -<br>h-caldesmone -<br>Vimentine +++ | Desmin -<br>aSMA -<br>h-caldesmone -<br>Vimentine +++ |
| SMC appearance | Organised in bundles | Scattered | No SMCs |
| Myofibroblasts | +++ | + | - |
| Associated Pathology | Inguinal Hernia (boys and girls) | Hydrocele | Obliterated PV<br>No pathology |

The absence, at a certain degree, of dedifferentiation has been suggested to be involved in the pathogenesis of childhood inguinoscrotal pathologies. The presence of myofibroblasts, cells that share morphological features with both fibroblasts and SMCs, were commonly encountered in sacs associated with inguinal hernia, but only infrequently in sacs of boys with undescended testis or hydrocele.

The presence of myofibroblasts seems to reflect attempted apoptosis by smooth muscle through dedifferentiation into an earlier stage, which appears to be an essential step for the obliteration of the PV (Table 3) [12, 24].

## CONCLUSION

Our findings support the theory proposed by Tanyel according to which SMCs that appear in the PV fail to complete their natural course of apoptosis. This fact impedes the process of PV obliteration at varying degrees, causing IH or hydrocele.

## REFERENCES

[1] Momoh JT. Obliteration of processus vaginalis and inguinal hernial sacs in children. *Can J Surg*. 1982 Sep;25(5):483-5.

[2] Barteczko KJ, Jacob MI. The testicular descent in human. Origin, development and fate of the GT Hunteri, processus vaginalis peritonei, and gonadal ligaments. *Adv. Anat Embryol Cell Biol*. 2000;156:III-X, 1-98.

[3] Hutson JM, Nation T, Balic A, Southwell BR. The role of the gubernaculum in the descent and undescent of the testis. *Ther Adv. Urol.* 2009 Jun;1(2):115-21.

[4] Attah AA, Hutson JM. The role of intra-abdominal pressure in cryptorchidism. *J. Urol.* 1993 Sep;150(3):994-6.

[5] Clarnette TD, Lam SK, Hutson JM. Ventriculo-peritoneal shunts in children reveal the natural history of closure of the processus vaginalis. *J. Pediatr. Surg*. 1998 Mar;33(3):413-6.

[6] Fiegel HC, Rolle U, Metzger R, Geyer C, Till H, Kluth D. The testicular descent in the rat: a scanning electron microscopic study. *Pediatr. Surg. Int.* 2010 Jun;26(6):643-7.

[7] Shrock P. The processus vaginalis and gubernaculums testis. Their raison d' etre redefined. *Surg. Clin. North Am*. 1971;51:1263-8.

[8] Gier HM, Marion GB. Development of mammalian testes and genital ducts. *Biol. Reprod.* 1969;1:1-23.

[9] Backhouse KM. The natural history of testicular descent and maldescent. *Proc. R Soc. Med.* 1966;59:357-60.

[10] Ramasamy M, Di Pilla N, Yap T, et al. Enlargement of the processus vaginalis during testicular descent in rats. *Pediatt. Surg. Int.* 2001;17:312-5.

[11] Buraundi S, Balic A, Farmer PJ, Southwell BR, Hutson JM. Gubernacular development in the mouse is similar to the rat and suggests that the PV is derived from the urogenital ridge and is different from the parietal peritoneum. *J. Pediatr. Surg.* 2011 Sep;46(9):1804-12.

[12] Mouravas V. [Histologic and immunohistochemical study of the hernia sac in children with malformations of the processus vaginalis]. Doctorate dissertation No 2168 , University of Thessaloniki Medical School, 2008.

[13] Tanyel FC, Dağdeviren A, Müftüoğlu S, Gürsoy MH, Yürüker S, Büyükpamukçu N. Inguinal hernia revisited through comparative evaluation of peritoneum, processus vaginalis, and sacs obtained from children with hernia, hydrocele, and undescended testis. *J. Pediatr. Surg.* 1999 Apr;34(4):552-5.

[14] Tanyel FC, Talim B, Atilla P, Müftüoğlu S, Kale G. Myogenesis within the human GT: histological and immunohistochemical evaluation. *Eur. J. Pediatr. Surg.* 2005 Jun;15(3):175-9.

[15] Costa WS, Francisco JB Sampaio, Luciano A, Favorito, Luiz EM Cardoso Testicular migration: remodeling of connective tissue and muscle cells in human GT testis. *J. Urol.* 2002 May ;167 (5):2171-6.

[16] Peacock EE Jr, Madden JW. Studies on the biology and treatment of recurrent inguinal hernia: II. Morphological changes. *Ann. Surg.* 1974; 179:567-571.

[17] Jain V, Srivastava R, Jha S, Misra S, Rawat NS, Amla DV. Study of Matrix Metalloproteinase-2 in IH. *J. Clin. Med. Res.* 2009 Dec;1(5): 285-9.

[18] Friedman DW, Boyd CD, Norton P, Greco RS, Boyarsky AH, Mackenzie JW, Deak SB. Increases in type III collagen gene expression and protein synthesis in patients with inguinal hernias. *Ann. Surg.* 1993;218(6):754-760.

[19] Hosgor M, Karaca I, Ozer E, Suzek D, Ulukus C, Ozdamar A. Do alterations in collagen synthesis play an etiologic role in childhood inguinoscrotal pathologies: an immunohistochemical study. *J. Pediatr. Surg.* 2004 Jul;39(7):1024-9.

[20] Cook BJ, Hasthorpe S, Hutson JM Fusion of childhood IH induced by HGF and CGRP via an epithelial transition. *J. Pediatr. Surg.* 2000 Jan;35(1):77-81.

[21] Hutson JM, Albano FR, Paxton G, Sugita Y, Connor R, Clarnette TD, Gray AZ, Watts LM, Farmer PJ, Hasthorpe S. In vitro fusion of human inguinal hernia with associated epithelial transformation. *Cells Tissues Organs* 166: 249–258.

[22] Tanyel FC, Müftüoglu S, Dagdeviren A, Kaymaz FF, Büyükpamukçu N Myofibroblasts defined by electron microscopy suggest the dedifferentiation of smooth muscle within the sac walls associated with congenital inguinal hernia. *BJU Int.* 2001 Feb;87(3):251-5.

[23] Tanyel FC, Talim B, Atilla P, Müftüoğlu S, Kale G. Myogenesis within the human gumbernaculum: histological and immunohistochemical evaluation. *Eur. J. Pediatr. Surg.* 2005 Jun;15(3):175-9.

[24] Mouravas VK, Koletsa T, Sfougaris DK, Philippopoulos A, Petropoulos AS, Zavitsanakis A, Kostopoulos I. Smooth muscle cell differentiation in the processus vaginalis of children with hernia or hydrocele Hernia. 2010 Apr;14(2):187-91.

[25] Tanyel FC, Talim B, Kale G, Büyükpamukçu N. Differences in the morphology of the processus vaginalis with sex and underlying disease condition. *Pathol. Res. Pract.* 2000;196(11):767-70.

[26] Tanyel FC, Erdem S, Büyükpamukçu N, Tan E. Smooth muscle within incomplete obliterations of processus vaginalis lacks apoptotic nuclei. *Urol. Int.* 2002;69(1):42-5.

[27] Tanyel FC. Obliteration of processus vaginalis: aberrations in the regulatory mechanism result in an inguinal hernia, hydrocele or undescended testis.*Turk. J. Pediatr.* 2004;46 Suppl:18-27.

[28] Beasley SW, Hutson JM. Effect of division of genitofemoral nerve on testicular descent in the rat. *Aust. N Z J Surg.* 1987 Jan;57(1):49-51.

[29] Clarnette T. D.,Hutson J. M. *The development and closure of the processus vaginalis Hernia* June 1999, Volume 3, Issue 2, pp 97-102.

[30] Schwindt B, Farmer PJ, Watts LM, Hrabovszky Z, Hutson JM Localization of calcitonin gene-related peptide within the genitofemoral nerve in immature rats. *J. Pediatr. Surg.* 1999 Jun;34(6):986-91.

[31] Hutson JM, Sasaki Y, Huynh J, Yong E, Ting A. The gubernaculum in testicular descent and cryptorchidism. *Turk. J. Pediatr.* 2004;46 Suppl: 3-6.

In: Muscle Cells
Editor: Benigno Pezzo

ISBN: 978-1-62417-233-5


*Chapter 4*

# KEY RESIDUES CAUSE DIFFERENTIAL GALLBLADDER RESPONSE TO PACAP AND VIP IN THE GUINEA PIG

***Muxin Wei[1], Yaofu Fan[1], Satoru Naruse[2], Kiyoshi Nokihara[7], Victor Wray[6], Tsuyoshi Ozaki[4], Eiji Ando[5], Kotoyo Fujiki[3,4], and Hiroshi Ishiguro[3]***

[1]The First Affiliated Hospital of Nanjing Medical University, Nanjing, China
[2]Miyoshi Municipal Hospital, Miyoshi, Japan
[3]Human Nutrition, Nagoya University Graduate School of Medicine, Nagoya, Japan
[4]National Institute of Physiological Sciences, Okazaki, Japan
[5]Biotechnology Instruments Department, Shimadzu Corporation, Kyoto, Japan
[6]Department of Structural Biology, Helmholtz Centre for Infection Research, Braunschweig, Germany
[7]HiPep Laboratories, Kyoto, Japan

## AIM

To investigate the effects of pituitary adenylate cyclase activating polypeptide (PACAP) and vasoactive intestinal peptide (VIP) in the guinea pig gallbladder,and identify key residues responsible for their interactions with PACAP (PAC1) and VIP (VPAC) receptors in the guinea pig gallbladder.

## METHODS

We synthesized the PACAP/VIP hybrid peptides by a simultaneous multiple solid-phase peptide synthesizer using the Fmoc strategy. The peptides were tested on the isolated guinea pig gallbladder using an improved horizontal-type organ bath.We extracted total RNA from the guinea pig gallbladder, RT-PCR was conducted using the primers with high sequence homology among human, mouse, and rat PAC1, VPAC1, VPAC2 receptors.

## RESULTS

VIP induced relaxation of gallbladder smooth muscle strips, while PACAP27 contracted them. Positions 4, 5, 9 and 24 26 can be replaced without significant loss in activity. [ $Leua^{13}$ ]-PACAP27, a substitution in the α-helix domain, also had no significant loss in activity ($P$<0.05). It was more potent than[$Gly^{8}$]- and [ $DAsp^{8}$]-PACAP27 and could substitute peptides at position 21. Des-[His1] and [ Ala6]-PACAP27 had no activity at [10-7]mol/L. [Gly8]-, [DAsp8]-, [Phe21]- and [Pro21] -PACAP27 at 10-7mol/L were about 25% of PACAP27 at 10a-7aa mol/L (P<0.05). In our previous studies, the N-terminus from position I to 8 showed no defined helix or strand structure. Substitution of PACAP in this region showed less potency than substitutions in other regions. [Ala4]- and [Val5]PACAP-27 were more potent than PACAP-27 in stimulating the gallbladder. In contrast, [Ala4, Val5]- and[Ala4, Val5, Asn9]PACAP-27 induced relaxation similarly to VIP.[Asn9]-, [Thr11]-, or [Leu13]PACAP-27 had 20 –70% contractile activity of PACAP-27, whereas [Asn24,Ser25,Ile26]PACAP-27 showed no change in the activity. All VIP analogs, including [Gly4,Ile5,Ser9]VIP, induced relaxation. In the presence of a PAC1 receptor antagonist, PACAP(6 –38),the contractile response to PACAP-27 was inhibited and relaxation became evident. RT-PCR analysis revealed abundant expressions of PAC1 receptor, “hop” splice variant, and VPAC1 and VPAC2 receptor mRNAs in the guinea pig gallbladder.PACAP-38 and PACAP-27 evoke opposite responses in the guinea pig gallbladder smooth muscle, where PACAP induces contraction while VIP causes relaxation. In addition the response to PACAP-38 is four times lower than that of PACAP-27.

## CONCLUSION

In conclusion, for the physiological action of PACAP in guinea pig gallbladder, the N-terminal disordered region is more important than other region. The disordered region from 1 to 8 is very important for physiological action. Position 21 is also important, however, because at a higher dose (13x10-7M) there was no significant loss in activity.The expression of the hop variant of PAC1 receptor may be related to the contractile response observed in the gallbladder. The effects caused by residues within the C-terminus are not a result of a response via the M-receptor or Na+ channel, but most likely arise from a delicate balance between the differential effects of PACAP-38 on specific PAC1 and VPACs receptors. PACAP-27 induces contraction of the gallbladder via PAC1/hop receptors. Gly4 and Ile5 are the key NH2-terminal residues of PACAP-27 that distinguish PAC1/hop receptors from VPAC1/VPAC2 receptors. However, both the NH2-ter-minal and -helical regions of PACAP-27 are required for initiating gallbladder contraction.

Pituitary adenylate cyclase-activating polypeptide (PACAP) is a member of the secretin/glucagon/vasoactive intestinal polypeptide (VIP) family of peptides. PACAP and VIP have opposite actions on the gallbladder ;PACAP induces contraction, whereas VIP induces relaxation.

Here, we have attempted to identify key residues responsible for their interactions with PACAP (PAC1) and VIP (VPAC) receptors in the guinea pig gallbladder.

PACAP-27 has a 68% sequence homology to VIP, and all are expressed in the central as well as peripheral and enteric nervous systems [1,2]. They are released from nerve terminals as neurotransmitters or neuromodulators and regulate the function of the brain and peripheral organs. PACAP exhibits protean biological effects on the gastrointestinal tract, including motility, secretion, and blood flow [3]. PACAP and VIP are coexpressed in nerve fibers and neurons in the ganglia of the guinea pig gallbladder [4].

Three receptor subtypes that recognize PACAP and VIP have been identified [2,5] (4, 22), and all belong to the group of seven transmembrane G protein-coupled receptors. The PACAP-specific ($PAC_1$) receptor has a much higher affinity for PACAP than VIP, whereas the classical VIP ($VPAC_1$) receptor and $VPAC_2$ receptor exhibit similar affinities for PACAP and VIP.

$VPAC_1$ and $VPAC_2$ receptors lead to activation of the adenylate cyclase/cAMP pathway in which elevation of intracellular cAMP, together with nitric oxide, mediates relaxation of intestinal and vascular smooth muscle cells [2,3]. $PAC_1$receptors, on the other hand, can activate the dual-signal transduction pathways involving adenylate cyclase and phospholipase C. The activation of the latter probably leads to inositol trisphosphate (IP3)-mediated $Ca^{2+}$ mobilization and protein kinase C mediated-gallbladder contraction [6], although until now it was not known which receptor subtype is expressed in the gallbladder.

PACAP and related peptides, except helodermin, show no stable structures in aqueous solution [7,8]. However, in more hydrophobic environments, i.e., in30-50% trifluoroethanol,

PACAP-38 has a stable structure consisting of three well defined domains:an initial disordered $NH_2$-terminus of eight residues, a central a-helical region from $Ser^9$ to $Val^{26}$ with a break between $Lys^{20}$ and $Lys^{21}$, and a COOH-terminal region with a short a-helix between $Gly^{28}$ and $Arg^{34}$ [8]. The structures of PACAP-27 and VIP resemble closely that of PACAP-38 except for the COOH-terminal region. The two helical structures of VIP involve residues $Thr^7$-$Lys^{15}$ and $Val^{19}$-$Leu^{27}$, and a flexible region exists between $Glu^{16}$ and $AIa^{18}$ [9]. These structural features define specific spatial arrangements of charged residues when they interact with their receptors. Because PACAP-27 has a 68% sequence homology to VIP, the difference in their interaction with PAC, receptors must reside in the nine amino acid residues that differentiate the two peptides (Table 1). In this study, using guinea pig gallbladder smooth muscle strips, to identify key residues for interaction of PACAP with $PAC_1$, receptors by exchanging amino acid residues of PACAP-27 with those of VIP and vice versa.

In the present study, we used isotonic transducers, which, by choosing an appropriate weight load, allowed us simultaneous measurements of both contractile and relaxant activities of the gallbladder smooth muscle strips. The stimulatory effect of PACAP-27 was independent of cholinergic nerves [10] and CCK (this study), the two major regulatory mechanisms of the gallbladder motility, but was significantly inhibited by PACAP, a PACAP-receptor antagonist. he relaxant effect of a high concentration of PACAP-27 ($10^{-6}$ M) can be partially blocked by a PACAP-receptor antagonist as well as by a VIP-receptor antagonist [10]. Taken together, it appears that PACAP-27 induces the gallbladder contraction directly via PACAP receptors and the relaxation via both PACAP receptors and PACAP/VIP common receptors.

Abundant expressions of $PAC_1$, $VPAC_1$,and $VPAC_2$ receptor mRNAs in the guinea pig gallbladder support this interpretation.

CCK mechanisms are activated, the inhibitory effects of PACAP and VIP via $VPAC_1$,and $VPAC_2$ receptors may be counterbalanced by the stimulatory effect via $PAC_1$ receptors. At higher concentrations($>3 \times 10^{-8}$ M), the activation of the stimulatory pathway via $PAC_1$ receptors overcomes the inhibitory one via $VPAC_1$ and $VPAC_2$ receptors, resulting in a biphasic response. Under a sustained contraction, circulation of the gallbladder may be maintained by vasodilator actions of PACAP and VIP via both $PAC_1$ and $VPAC_1$ and $VPAC_2$ receptors on blood vessels.

**Table 1. Amino Acid Sequence of PACAP-27, VIP, and Their Analogs**

PACAP-27 HSDGIFTDSY SRYRKQMAVK KYLAAVL-NH2
VIP HSDAVFTDNY TRLRKQMAVK KYLNSILN-NH2
[des-His1]PACAP-27 □SDGIFTDSY SRYRKQMAVK KYLAAVL-NH2
[Ala4]PACAP-27 HSDAIFTDSY SRYRKQMAVK KYLAAVL-NH2
[Val5]PACAP-27 HSDGVFTDSY SRYRKQMAVK KYLAAVL-NH2
[Ala4, Val5]PACAP-27 HSDAVFTDSY SRYRKQMAVK KYLAAVL-NH2
[Ala4, Val5, Asn9]PACAP-27 HSDAVFTDNY SRYRKQMAVK KYLAAVL-NH2
[Ala6]PACAP-27 HSDGIATDSY SRYRKQMAVK KYLAAVL-NH2
[DAsp8]PACAP-27 HSDGIFTDSY SRYRKQMAVK KYLAAVL-NH2
[Gly8]PACAP-27 HSDGIFTGSY SRYRKQMAVK KYLAAVL-NH2
[Asn9]PACAP-27 HSDGIFTDNY SRYRKQMAVK KYLAAVL-NH2
[Thr11]PACAP-27 HSDGIFTDSY TRYRKQMAVK KYLAAVL-NH2
[Leu13]PACAP-27 HSDGIFTDSY SRLRKQMAVK KYLAAVL-NH2
[Ala21]PACAP-27 HSDGIFTDSY SRYRKQMAVK AYLAAVL-NH2
[Phe21]PACAPAVK HSDGIFTDSY SRYRKQMAVK FYLAAVL-NH2
[Pro21]PACAP-27 HSDGIFTDSY SRYRKQMAVK PYLAAVL-NH2
[Asn24, Ser25, IIe26]PACAP-27 HSDGIFTDSY SRYRKQMAVK KYLNSIL-NH2
[Gly4]VIP HSDGVFTDNY TRLRKQMAVK KYLNSILN-NH2
[Ile5]VIP HSDAIFTDNY TRLRKQMAVK KYLNSILN-NH2
[Gly4, Ile5]VIP HSDGIFTDNY TRLRKQMAVK KYLNSILN-NH2
[Gly4, Ile5, Ser9]VIP HSDGIFTDSY TRLRKQMAVK KYLNSILN-NH2
[Ser9]VIP HSDAVFTDSY TRLRKQMAVK KYLNSILN-NH2
[Ser11]VIP HSDAVFTDNY SRLRKQMAVK KYLNSILN-NH2
[Tyr13]VIP HSDAVFTDNY TRYRKQMAVK KYLNSILN-NH2

The present study demonstrates the expression of an isoform of the $PAC_1$ receptor in the guinea pig gallbladder. The nucleotide sequence analysis revealed that the isoform was a splice variant that contained an additional 84 nucleotides encoding 28 amino acids in the third intracellular loop, the key domain for coupling to phospholipase C via a specific G protein. The deduced amino acid sequence was identical to that of a "hop" variant reported in rats [11] and humans [12]. Alternative splicing of two exons of rat PAC1 receptor gene generates four major splice variants, named hip, hop1, hop2, and hip-hop2. Each splice variant can be differentially coupled to two intracellular signal transduction pathways and thus results in variable elevations of cAMP and IP3 in a tissue specific manner [11]. Among the four splice variants of human $PAC_1$ receptors, the hop variant had a fivefold greater efficacy in IP3 production than the authentic $PAC_1$ receptor [12].Because PACAP induces relaxation of smooth muscles in most of the tissues that express $PAC_1$ receptors [3], the expression of the hop variant of $PAC_1$ receptor might be related to the contractile response observed in the gallbladder. Further studies are necessary to identify the cellular localizations of $PAC_1$ receptors, the hop variant, and $VPAC_1$and $VPAC_2$ receptors in the gallbladder.

PACAP-27 induces contraction of the gallbladder smooth muscles via $PAC_1$ and/or its hop variant receptors. The positions 4 and 5 are the key NH2-terminal residues of PACAP-27 that distinguish $PAC_1$/hop receptors from $VPAC_1$/$VPAC_2$ receptors in the gallbladder. However, both the NH2-terminal disordered region and -helical region of PACAP-27 are required for initiating gallbladder contraction. Tissue-specific expressions of PACAP and VIP and their receptors determine the net functions of PACAP and VIP.

## Differences in Action of PACAP-27 and PACAP-38 on Guinea Pig Gallbladder Smooth Muscle

Pituitary adenylate cyclase-activating polypeptide (PACAP) is a member of the secretin/glucagon/vasoactive intestinal polypeptide (VIP) family of peptides. Two bioactive molecules, PACAP-38 and PACAP-27, have been isolated with identical N-terminal sequences and one of these has an 11-residue C-terminal elongation. PACAP exhibits several biological functions on the gastrointestinal tract, including motility,secretion and blood flow [3]. A comparison of the primary structures is shown in Fig. 2. In our previous studies we have demonstrated that PACAP-38 and PACAP-27 are potent VIP-like vasodilators of the femoral arterial bed in dogs, but PACAP-38 differs from PACAP-27 and VIP in its prolonged vasodilatory effects on femoral blood flow [13]. To clarify the difference between VIP and the two PACAPs, a mini-library of VIP-PACAP peptides, which included VIP-PACAP hybrid peptides,were constructed by solid-phase peptide synthesis [14,15,16]. In addition, the solution structures of the PACAPs and their environmental dependence have been elucidated by CD and NMR studies. The global features of VIP are the same as that of PACAP-2 [17], and both have the same features as the 27 N-terminal residues of PACAP-38. The difference between PACAP-27 and PACAP-38 is the addition of a further C terminal helical region between residues 28 and 34.

PACAP/VIP receptors in gallbladder has also been reported [10]. However, it is only recently that the key residues, responsible for the interactions with their receptors PAC1 and VPAC, has been clarified through receptor mRNA studies [18]. The substitution of positions 4 and 5, [$Ala^4$, $Val^5$]PACAP-27 is different to other PACAP analogues, in that this exchange causes relaxation of the gallbladder. Position 6 is also important for contraction. In the present report we focus our attention on the action on the isolated gallbladder of PACAP-38,especially the significance of the C-terminal elongated segment of PACAP-38 using a synthetic mini-library of C-terminally deleted peptides.

The effects of VIP and PACAP on the guinea pig gallbladder smooth muscle strips were first reported in 1994, where VIP induced relaxation while PACAPs induced contraction in a concentration-dependent manner [9]. As the difference between PACAP-27 and -38 is the 11-residue C-terminal elongation of the latter. C-terminally deleted PACAP-38 peptides were synthesized and tested using PACAP-27 as control. All peptides evoked a dose dependent response in the gall bladder and in all cases, where meaningful responses could be measured, the PACAP-38 fragments showed a diminished response compared to PACAP-27.

It is known that PACAP-27 causes a biphasic response in gall bladder smooth muscle that corresponds to an initial large contractile response followed by a very weak relaxation. It is also known that guinea pig gallbladder expresses PAC1, VPAC1 and VPAC2 receptors and additionally other receptors that govern motility such as the muscarinic- and CCK-receptors. The effects of atropine, an antagonist of the PAC1, VPAC1 and VPAC2 receptors, and TTX, an inhibitor of the muscarinic- and CCK-receptors, only partially inhibited PACAP-38 contraction and their effects were not distinguishable. The dose-dependent decrease in gallbladder contraction caused by PACAP-38 is unambiguous evidence that the gallbladder posses PAC1 receptors. Careful observation showed that after inhibition of PACAP-38-contraction by PACAP(6-38), the gallbladder slowly relaxed. PACAP-38 also binds weekly to VPAC1 and VPAC2 receptors that cause a delicate balance between the relaxation effects of the VPAC receptors and the counterbalancing stimulatory effect of the PAC1 receptors.

## REFERENCES

[1] Arimura A. Pituitary adenylate cyclase activating polypeptide (PACAP):discovery and current status of research. *Regul. Pept.* 37: 287–303, 1992.

[2] Vaudry D., Gonzalez B. J., Basille M., Yon L., Fournier A., Vaudry H.Pituitary adenylate cyclase-activating polypeptide and its receptors: from structure to functions. *Pharmacol. Rev.* 52: 269 –324, 2000.

[3] Lauff J. M., Modlin I. M., Tang L. H. Biological relevance of pituitary adenylate cyclase-activating polypeptide (P. A. C. A. P.) in the gastrointestinal tract. *Regul. Pept.* 84: 1–12, 1999.

[4] Mawe G. M., Ellis L. M. Chemical coding of intrinsic and extrinsic nerves in the guinea pig gallbladder: distributions of P. A. C. A. P. and orphanin F. Q. *Anat. Rec.* 262: 101–109, 2001.

[5] Harmar A. J., Arimura A., Gozes I., Journot L., Laburthe M., Pisegna J. R.,Rawlings S. R., Robberecht P., Said S. I., Sreedharan S. P., Wank S. A.,Waschek J. A. International Union of Pharmacology. XVIII. Nomenclature of receptors for vasoactive intestinal peptide and pituitary adenylate cyclase-activating polypeptide. *Pharmacol. Rev.* 50: 265–270, 1998.

[6] Pang P. K., Kline L. W. Protein kinase C mediates the contractile actions of pituitary adenylate cyclase activating polypeptide in guinea pig gallbladder strips. *Regul. Pept.* 77:63– 67, 1998.

[7] Blankenfeldt W., Nokihara K., Naruse S., Lessel U., Schomburg D.,Wray V., N. M. R. spectroscopic evidence that helodermin, unlike othermembers of the secretin/VIP family of peptides, is substantially structured in water. *Biochemistry* 35: 5955–5962, 1996.

[8] Wray V., Kakoschke C., Nokihara K., Naruse S. Solution structure of pituitary adenylate cyclase activating polypeptide by nuclear magnetic resonance spectroscopy. *Biochemistry* 32: 5832–5841, 1993.

[9] Theriault Y., Boulanger Y., St-Pierre S. Structural determination of the vasoactive intestinal peptide by two-dimensional H-NMR spectroscopy. *Biopolymers* 31: 459 – 464, 1991.

[10] Parkman H. P., Pagano A. P., Ryan J. P. Dual effects of P. A. C. A. P. on guinea pig gallbladder muscle via PACAP-preferring and VIP/PACAP-preferring receptors. *Am. J. Physiol. Gastrointest. Liver Physiol.* 272: G1433–G1438,1997.

[11] Spengler D., Waeber C., Pantaloni C., Holsboer F., Bockaert J., Seeburg P. H., Journot L. Differential signal transduction by five splice variants of the P. A. C. A. P. receptor. *Nature* 365: 170 –175, 1993.

[12] Pisegna J. R., Wank S. A. Cloning and characterization of the signal transduction of four splice variants of the human pituitary adenylate cyclase activating polypeptide receptor. Evidence for dual coupling to adenylate cyclase and phospholipase C. *J. Biol. Chem.* 271: 17267–17274,1996.

[13] Naruse S., Suzuki T., Ozaki T., Nokihara K. Vasodilator effect of pituitary adenylate cyclase activating polypeptide (P. A. C. A. P.) on femoral blood flow in dogs. *Peptides* 14:505–510.

[14] Nokihara K., Naruse S., Ando E., Wray V. Synthesis and structure-activity relationship of PACAP-VIP related peptides.In: Schneider C. H., Eberle A. N. (eds) *Peptides* 1992. E. S. C. O. M. Science Publishers B. V., Leiden, pp. 723–724.

[15] Nokihara K., Ando E., Naruse S., Nakamura T., Wray V. Highly efficient synthesis of P. A. C. A. P. and its related peptides and their biological actions in conscious dogs. In: Okada Y (ed) *Peptide Chemistry* 1993. Protein Research Foundation, Osaka, pp. 265–268.

[16] Nokihara K., Ando E., Naruse S., Wei M., Wray V. Synthesis and structure-activity relationship of VIP-PACAP hybrid peptides. In: Ohno M (ed) *Peptide Chemistry* 1994. Protein Research Foundation, Osaka, pp. 53–56.

[17] Fry D. C., Madison V. S., Bolin D. R., Greeley D. N., Toome V., Wegrzynski B. B. Solution structure of an analogue of vasoactive intestinal peptide as determined by two-dimensional N. M. R. and circular dichroism spectroscopies and constrained molecular dynamics. *Biochemistry* 28:2399–2409.

[18] Wei M., Fujiki K., Ando E., Zhang S., Ozaki T., Ishiguro H., Kondo T.,Nokihara K., Wray V., Naruse S. . Identification of key residues that cause differential gallbladder response to P. A. C. A. P. and V. I. P. in the guinea pig. *Am. J. Physiol.* 292:G76–G83.

[19] Wei M., Naruse S., Nakamura T., Nokihara K., Ozaki T. The effect of pituitary adenylate cyclase activation polypeptide (P. A. C. A. P.) 38 on gallbladder smooth muscle in vitro. *Biomed. Res.* 15:221–223.

In: Muscle Cells
Editor: Benigno Pezzo

ISBN: 978-1-62417-233-5


*Chapter 5*

# MUSCULAR OPTOGENETICS: CONTROLLING MUSCLE FUNCTIONS WITH LIGHT

***Toshifumi Asano[1,2], Toru Ishizuka[1,3] and Hiromu Yawo[1,3,4]***

[1]Department of Developmental Biology and Neuroscience, Tohoku University Graduate School of Life Sciences, Sendai, Japan

[2]Japan Society for the Promotion of Science, Tokyo, Japan

[3]Japan Science and Technology Agency (JST), Core Research of Evolutional Science & Technology (CREST), Tokyo, Japan

[4]Center for Neuroscience, Tohoku University Graduate School of Medicine, Sendai, Japan

## ABSTRACT

Traditionally, artificial contractions of muscles have been induced electrically, mechanically or pharmacologically to investigated their functional characteristics. Although simple and convenient, these techniques are generally non-specific, non-uniform and invasive. To improve the spatiotemporal resolution and to reduce the invasiveness, the optogenetic approach using light-sensitive proteins has attracted attention as a new method. Recent examples include using channelrhodopsin-2 (ChR2), a light-activated ion channel from a green alga, for optical pacing of cardiomyocytes, the optical control of C2C12 myoblast-derived myotubes and the optically induced maturation of cultured myotubes. The optical manipulation of muscle activities would facilitate *in vitro* studies of muscle contraction through manipulating/modulating specific biological processes during myogenic development. It has potential therapeutic applications for producing light-sensitive human muscle substitutes for muscle weakness such as muscular dystrophy and amyotrophic lateral sclerosis (ALS). It could also enable the development of a wireless driving source of muscle-powered actuators/microdevices. Here, this chapter reviews a general overview of the state of research and future prospects and challenges of optogenetics for muscle cells.

## INTRODUCTION

Muscle activity plays important roles in health and contributes to the prevention and improvement of many chronic diseases, such as obesity, type 2 diabetes, sarcopenia, neurodegeneration and osteoporosis [1, 2]. Muscle contraction induces the gene expression and protein synthesis of a number of molecules such as acetylcholinesterase [3], L-type $Ca^{2+}$ channels [4] and glucose transporter 4 [5]. The cellular mechanisms underlying these reactions have been investigated with an *in vitro* model system using cultured myotubes. To evoke the contraction and the subsequent biochemical reactions, the myotubes *in vitro* were stimulated either electrically [4, 6], mechanically [7] or pharmacologically [8]. For example, electrical field stimulation (EFS) during the period of myogenic development facilitates the maturation of skeletal myotubes such as the assembly of the sarcomere, the smallest contractile units in striated muscle, elicits the transient fluctuation of intracellular $Ca^{2+}$ following an EFS [9], and enhances the contractile properties together with the expression of elongation factors and muscle proteins [6, 10-13]. Although, EFS is a simple and common method for applying short-term, patterned excitation of a cell, its effect is often nonuniform and many untargeted myotubes are stimulated simultaneously. As metal electrodes are placed in the extracellular space during EFS, the long-term stimulation inevitably has undesirable effects on the cell because of the production of toxic gases, such as $H_2$ and $Cl_2$, and alterations of the pH due to Faradaic reaction. Therefore, the magnitude and duration of the EFS is limited to the range of a few volts/mm and milliseconds, respectively.

Recently, an optogenetic approach using light-sensitive ion channels from a green alga *Chlamydomonas reinhardtii*, channelrhodopsins (ChRs), has attracted much attention as a new method to overcome the above limits of EFS [14-17]. Each ChR is a member of the microbial-type (archaeal-type, type I) rhodopsin family with a core structure of about 300 amino acids. The core structure consists of seven transmembrane domains (TM1-7) and a retinal that is covalently bound to the conserved Lys residue at the middle of TM7. Light absorption is followed by the photoisomerization of the all-*trans* retinal to a 13-*cis* configuration and subsequent conformational changes of the molecule, which allow the channel structure to become permeable to cations, such as $Na^+$, $K^+$, $Ca^{2+}$ and $H^+$ [18-20]. This enables very rapid (in the orders of ms) generation of an inward current in the cells expressing ChRs and induces membrane depolarization [21-24]. The light-induced depolarization of ChR-expressing neurons generates action potentials through the activation of voltage-gated ion channels. This method has the obvious advantages of fine spatial and temporal resolution, ability for parallel stimulations at multiple sites, and relative harmlessness and convenience. This chapter provides a general overview and describes prospective applications of optogenetic techniques focusing on muscle cells.

## Optogenetic Stimulation of Cardiac Cells

Although the application of optogenetics is expanding in the field of neuroscience and revealing the functional connections of neurons in the brain, it is still limited in other excitable cells such as cardiac, smooth and skeletal muscles. Embryonic stem (ES) cells were transfected with one of the ChR2 variants, ChR2(H134R), and differentiated to become light-sensitive cardiomyocytes *in vitro* capable of optical modification in terms of their pacemaking activities. For example, optical stimulation caused the prolongation of depolarization with enhanced $Ca^{2+}$ influx in ES-derived cardiomyocytes [25]. The electrical and mechanical activities of ES-derived cardiomyocytes could be paced to follow the light stimulation [25, 26]. Based on these *in vitro* experiments, light-induced pacing of the human heart could be explored by computer simulation. Indeed, atrial irradiation prolonged both the P-wave duration and the PQ interval of the electrocardiogram (ECG), whereas ventricular irradiation affected the QRS duration in the ChR2-expressing heart of transgenic mice [25, 26]. Using transgenic zebrafish that stably express ChR(H134R) or halorhodopsin (NpHR), a light-driven chloride ion pump, in the cardiomyocytes, the cardiac pacemaking region was mapped by spatial patterns of irradiation generated by a digital micromirror device [27]. Irradiation of the genetically engineered fish heart could modify the heart rate and switch the heartbeat from a healthy to diseased state depending on the light-induced excitation or inhibition. On the other hand, optogenetic pacing using a nonviral strategy could be achieved in cardiomyocytes that had been co-cultured with human embryonic kidney (HEK) 293 cells expressing ChR2(H134R) to form a syncytium [28]. The irradiation of blue LED light generated an inward current in the HEK293 cells, resulting in depolarization of the nearby cardiac tissues through the gap junction channels of connexin 43 (Cx43) and, eventually, inducing propagating action potentials. This approach may yield not only new avenues for basic studies of cardiac arrhythmias *in vitro*, but also potential clinical applications of the optical pacemaker as a low energy substitute for conventional cardiac pacemakers.

## Optical Manipulation of Skeletal Muscle Functions

Skeletal muscle has a unique mechanism of excitation-contraction (E-C) coupling to initiate contraction with high responsiveness and accuracy [29-31]. In terms of the vertebrate skeletal muscle, motor neuron activity is transmitted across the neuromuscular junction to depolarize a myocyte, which generates an action potential. The action potential is conducted along the surface membrane and depolarizes the transverse tubular membrane. This depolarization is then sensed by dihydropyridine receptors (DHPRs), α-subunits of voltage-dependent L-type calcium channels, which couple with ryanodine receptors (RyRs) in the sarcoplasmic reticulum (SR) to liberate $Ca^{2+}$ from the intracellular $Ca^{2+}$ stores. The resultant increase of intracellular $Ca^{2+}$ triggers the vigorous contraction of the myofibrils.

Recently, Asano *et al.* [32] produced photosensitive skeletal muscle cells *in vitro* from C2C12 myoblasts, an immortal cell line of murine skeletal myoblasts originally derived from satellite cells [33], into which the ChR2 gene was introduced using a retroviral vector. The cloned ChR2-expressing C2C12 myoblasts were fused with unrecombinant C2C12 to form multinucleated myotubes and allowed to become contractile mature muscle fibers. The ChR2-expressing muscle fibers were depolarized by a blue LED and eventually evoked action potentials in a manner dependent on the intensity and duration of the irradiation. This was followed by obvious contractions synchronous with the light pulses of the given temporal pattern, a twitch-like contraction at low frequency (1-4 Hz) and a tetanus-like contraction at high frequency (5-10 Hz). The optically evoked contractile responses were similar to those evoked electrically in terms of both the contractile pattern and the magnitude.

## OPTOGENETIC MATURATION OF CONTRACTILE MUSCLES

Although the cultured C2C12 myoblasts fuse with each other to form multinucleated myotubes after a few days in the differentiation medium, their maturation with contractile ability and a sarcomere structure was difficult to attain under conventional conditions. Previously, maturation was reported to be facilitated by extracellular EFS, with the appearance of the sarcomere assembly and contractile property [9]. It is thus hypothesized that light-evoked oscillation of the membrane depolarization combined with ChRs would accelerate the maturation of a skeletal myotube during myogenic differentiation. Indeed, when optical stimulation with short LED pulses was periodically applied to C2C12 myotubes that express a chimeric channelrhodopsin, channelrhodopsin-green receiver (ChRGR), the number of contractile myotubes was significantly increased. These myotubes had the characteristic striation pattern resulting from the regular alignment of sarcomeric proteins, Z-line protein α-actinin and A-band protein skeletal myosin heavy chain (MHC). In contrast, the non-stimulated control myotubes rarely showed the striated patterns with sarcomeric α-actinin and MHC, which were diffusely distributed in punctate patterns or localized along filamentous structures. These results suggest that optogenetics could be used to manipulate the myogenic contraction and the maturation of skeletal muscle cells.

## CONCLUSION

In this chapter, the present and potential progress of muscular optogenetics as a new technical approach to solve several problems in the conventional study of muscle cells has been described. The contractile activity and the myogenic development can be regulated by lightening patterns using optogenetics. The optically regulated pacing of cardiomyocytes could become a substitute for electrical pacemakers/defibrillators, which are surgically implanted devices with electrodes inserted into heart tissue. To overcome muscle weakness such as that resulting from muscular dystrophy and amyotrophic lateral sclerosis (ALS), human muscle tissue could also be replaced through optogenetically facilitated myogenic development of myoblasts derived from induced pluripotent stem (iPS) cells or mesenchymal stem cells derived from a recipient. The contraction of transplanted muscles could also be optically regulated with high accuracy and non-invasively without the need for electrodes. Skeletal muscle cells are a high performance force transducers that can generate contractile energy efficiently through biochemical reactions. With savings in energy, resources and space, the optical control of photosensitive muscles could extend the range of bioengineering applications, such as the generation of wireless driving devices equipped with muscle-powered actuators.

## References

[1] Handschin, C.; Spiegelman, B. M., The role of exercise and PGC1α in inflammation and chronic disease. *Nature* 2008, 454, (7203), 463-469.

[2] Booth, F. W.; Chakravarthy, M. V.; Gordon, S. E.; Spangenburg, E. E., Waging war on physical inactivity: using modern molecular ammunition against an ancient enemy. *Journal of Applied Physiology* 2002, 93, (1), 3-30.

[3] Sketelj, J.; Leisner, E.; Gohlsch, B.; Škorjanc, D.; Pette, D., Specific impulse patterns regulate acetylcholinesterase activity in skeletal muscles of rats and rabbits. *Journal of Neuroscience Research* 1997, 47, (1), 49-57.

[4] Freud-Silverberg, M.; Shainberg, A., Electric stimulation regulates the level of Ca-channels in chick muscle culture. *Neuroscience Letters* 1993, 151, (1), 104-106.

[5] Hofmann, S.; Pette, D., Low-frequency stimulation of rat fast-twitch muscle enhances the expression of hexokinase II and both the translocation and expression of glucose transporter 4 (GLUT-4). *European Journal of Biochemistry* 1994, 219, (1-2), 307-315.

[6] Burch, N.; Arnold, A. S.; Item, F.; Summermatter, S.; Brochmann Santana Santos, G.; Christe, M.; Boutellier, U.; Toigo, M.; Handschin, C., Electric pulse stimulation of cultured murine muscle cells reproduces gene expression changes of trained mouse muscle. *Plos One* 2010, 5, (6), e10970.

[7] De Deyne, P. G., Formation of sarcomeres in developing myotubes: role of mechanical stretch and contractile activation. *American Journal of Physiology: Cell Physiology* 2000, 279, (6), C1801-C1811.

[8] Nakanishi, K.; Dohmae, N.; Morishima, N., Endoplasmic reticulum stress increases myofiber formation *in vitro*. *FASEB Journal* 2007, 21, (11), 2994-3003.

[9] Fujita, H.; Nedachi, T.; Kanzaki, M., Accelerated *de novo* sarcomere assembly by electric pulse stimulation in C2C12 myotubes. *Experimental Cell Research* 2007, 313, (9), 1853-1865.

[10] Thelen, M. H.; Simonides, W. S.; van Hardeveld, C., Electrical stimulation of C2C12 myotubes induces contractions and represses thyroid-hormone-dependent transcription of the fast-type sarcoplasmic-reticulum $Ca^{2+}$-ATPase gene. *Biochemical Journal* 1997, 321, (3), 845-848.

[11] Park, H.; Bhalla, R.; Saigal, R.; Radisic, M.; Watson, N.; Langer, R.; Vunjak-Novakovic, G., Effects of electrical stimulation in C2C12 muscle constructs. *Journal of Tissue Engineering and Regenerative Medicine* 2008, 2, (5), 279-287.

[12] Stern-Straeter, J.; Bach, A. D.; Stangenberg, L.; Foerster, V. T.; Horch, R. E.; Stark, G. B.; Beier, J. P., Impact of electrical stimulation on three-dimensional myoblast cultures - a real-time RT-PCR study. *Journal of Cellular and Molecular Medicine* 2005, 9, (4), 883-892.

[13] Pedrotty, D. M.; Koh, J.; Davis, B. H.; Taylor, D. A.; Wolf, P.; Niklason, L. E., Engineering skeletal myoblasts: roles of three-dimensional culture and electrical stimulation. *American Journal of Physiology: Heart and Circulatory Physiology* 2005, 288, (4), H1620-H1626.

[14] Deisseroth, K.; Feng, G.; Majewska, A. K.; Miesenböck, G.; Ting, A.; Schnitzer, M. J., Next-generation optical technologies for illuminating genetically targeted brain circuits. *The Journal of Neuroscience* 2006, 26, (41), 10380-10386.

[15] Bernstein, J. G.; Boyden, E. S., Optogenetic tools for analyzing the neural circuits of behavior. *Trends in Cognitive Sciences* 2011, 15, (12), 592-600.

[16] Carter, M. E.; de Lecea, L., Optogenetic investigation of neural circuits *in vivo*. *Trends in Molecular Medicine* 2011, 17, (4), 197-206.

[17] Rein, M. L.; Deussing, J. M., The optogenetic (r)evolution. *Molecular Genetics and Genomics* 2012, 287, (2), 95-109.

[18] Bamann, C.; Kirsch, T.; Nagel, G.; Bamberg, E., Spectral characteristics of the photocycle of channelrhodopsin-2 and its implication for channel function. *Journal of Molecular Biology* 2008, 375, (3), 686-694.

[19] Ernst, O. P.; Sánchez Murcia, P. A.; Daldrop, P.; Tsunoda, S. P.; Kateriya, S.; Hegemann, P., Photoactivation of channelrhodopsin. *Journal of Biological Chemistry* 2008, 283, (3), 1637-1643.

[20] Stehfest, K.; Hegemann, P., Evolution of the channelrhodopsin photocycle model. *Chemphyschem* 2010, 11, (6), 1120-1126.

[21] Nagel, G.; Ollig, D.; Fuhrmann, M.; Kateriya, S.; Musti, A. M.; Bamberg, E.; Hegemann, P., Channelrhodopsin-1: a light-gated proton channel in green algae. *Science* 2002, 296, (5577), 2395-2398.

[22] Nagel, G.; Szellas, T.; Huhn, W.; Kateriya, S.; Adeishvili, N.; Berthold, P.; Ollig, D.; Hegemann, P.; Bamberg, E., Channelrhodopsin-2, a directly light-gated cation-selective membrane channel. *Proceedings of the National Academy of Sciences of the United States of America* 2003, 100, (24), 13940-13945.

[23] Boyden, E. S.; Zhang, F.; Bamberg, E.; Nagel, G.; Deisseroth, K., Millisecond-timescale, genetically targeted optical control of neural activity. *Nature Neuroscience* 2005, 8, (9), 1263-1268.

[24] Ishizuka, T.; Kakuda, M.; Araki, R.; Yawo, H., Kinetic evaluation of photosensitivity in genetically engineered neurons expressing green algae light-gated channels. *Neuroscience Research* 2006, 54, (2), 85-94.

[25] Bruegmann, T.; Malan, D.; Hesse, M.; Beiert, T.; Fuegemann, C. J.; Fleischmann, B. K.; Sasse, P., Optogenetic control of heart muscle *in vitro* and *in vivo*. *Nature Methods* 2010, 7, (11), 897-900.

[26] Abilez, O. J.; Wong, J.; Prakash, R.; Deisseroth, K.; Zarins, C. K.; Kuhl, E., Multiscale computational models for optogenetic control of cardiac function. *Biophysical Journal* 2011, 101, (6), 1326-1334.

[27] Arrenberg, A. B.; Stainier, D. Y.; Baier, H.; Huisken, J., Optogenetic control of cardiac function. *Science* 2010, 330, (6006), 971-974.

[28] Jia, Z.; Valiunas, V.; Lu, Z.; Bien, H.; Liu, H.; Wang, H.-Z.; Rosati, B.; Brink, P. R.; Cohen, I. S.; Entcheva, E., Stimulating cardiac muscle by light: cardiac optogenetics by cell delivery. *Circulation: Arrhythmia and Electrophysiology* 2011, 4, (5), 753-760.

[29] Ríos, E.; Pizarro, G., Voltage sensor of excitation-contraction coupling in skeletal muscle. *Physiological Reviews* 1991, 71, (3), 849-908.

[30] Lamb, G., Excitation–contraction coupling in skeletal muscle: comparisons with cardiac muscle. *Clinical and Experimental Pharmacology and Physiology* 2000, 27, (3), 216-224.

[31] Jurkat-Rott, K.; Lehmann-Horn, F., Muscle channelopathies and critical points in functional and genetic studies. *Journal of Clinical Investigation* 2005, 115, (8), 2000-2009.

[32] Asano, T.; Ishizuka, T.; Yawo, H., Optically controlled contraction of photosensitive skeletal muscle cells. *Biotechnology and Bioengineering* 2012, 109, (1), 199-204.

[33] Yaffe, D.; Saxel, O., Serial passaging and differentiation of myogenic cells isolated from dystrophic mouse muscle. *Nature* 1977, 270, (5639), 725-727.

# INDEX

## A



## B

## C

## D



## E



## F

## G



## H

## I



## K



## L



## M

## N

## O



## P

## Q



## R

## S



## T

## U



## V



## W

## Y



## Z